GRANT WRITING 101

GRANT WRITING 101

Everything You Need to Start Raising Funds Today

VICTORIA M. JOHNSON

New York Chicago San Francisco Lisbon London
Madrid Mexico City Milan New Delhi San Juan
Seoul Singapore Sydney Toronto

The McGraw-Hill Companies

1 2 3 4 5 6 7 8 9 10 QFR/QFR 1 0 9 8 7 6 5 4 3 2 1 0

ISBN 978-0-07-175018-9
MHID 0-07-175018-5

Library of Congress Cataloging-in-Publication Data

Johnson, Victoria M.
Grant writing 101 : everything you need to start raising funds today / by Victoria M. Johnson.
p. cm.
ISBN 978–0–07–175018–9 (alk. paper)
1. Proposal writing for grants. 2. Fund raising. I. Title.
HG177.J64 2011
658.15'224—dc22 2010039599

McGraw-Hill books are available at special quantity discounts to use as premiums and sales promotions or for use in corporate training programs. To contact a representative, please e-mail us at bulksales@mcgraw-hill.com.

This book is printed on acid-free paper.

This book would not have come to fruition without the support and love of my husband, Michael. I am happy (and lucky) to have you in my life.

Contents

Preface

As an experienced grant writer, friends ask me for advice for their organizations. I happily give it, but they need more. They need to learn how to write a grant themselves, they need to know the bare minimum to include in a proposal, they need to be shown what to do if their proposal is rejected, and they need tips for long-term success. But I'm only one person. How can I help so many? This book is for those friends and for all of you doing extraordinary work in the community with minimal staff and budgets and no time to take grant writing classes. This book is for the tens of thousands of nonprofit organizations and individuals who need a simple, straight-for-the-jugular but down-to-earth guide to grant writing.

Grant Writing 101: Everything You Need to Start Raising Funds Today gives frank advice and strategies for people who need to write a grant right now. There's no history of grant writing, no theories, no long-winded passages to discourage a stressed grant seeker. The book contains only kick-butt guerilla tactics that you can use immediately. In effect, only the essentials will be presented to you. The guerilla mind-set is the best line of attack for someone writing a grant for the first time. It's clear-cut, without the fluff or the tedious lectures.

A guerilla grant writer takes a positive course of action, seizes opportunities, and is well prepared for any grant prospect. Each chapter in Part 1 includes guerilla grant writing approaches that will improve your grant writing skills and a summary to review the material and give you a kick in the pants. You'll learn how to break down complex applications into manageable steps. You'll train to describe with compelling language. You'll build your arsenal of powerful sentences. Part 2 offers a variety of samples of grant proposals and other ammo to fortify your grant writing efforts.

Your organization changes lives, right? And you agree one person can change the world. As a grant writer, you'll find funds for your organization to do its work. In essence, this means that *you* can change the world—one grant at a time.

My goal is to empower you to write successful grants and to show you how grant writing will help you to grow your organization. Grant writing isn't hard. It is a basic process, and you are the best person in your organization to do it. You have this book as a guide, and it includes sample grant proposals and expert advice from industry professionals. Congratulations on taking this first step. You'll be writing kick-butt grant proposals in no time.

Acknowledgments

My gratitude goes to my brilliant agent, Paul S. Levine; to Michele R. Wells, the astute senior editor at McGraw-Hill who discovered this project; to Brian J. Foster, the awesome associate editor at McGraw-Hill Professional for his outstanding guidance, patience, and sharp eye; and to all the other amazing folks at McGraw-Hill whom I've never met but who've had a hand in making this book a reality.

I'd also like to thank Lita A. Kurth, a friend who happens to be a professional editor, for reading the draft of this book with her editor's pencil in hand.

A heart-felt thank you and deep appreciation goes to all the experts, colleagues, and friends for their contributions to this book. They generously provided their words of wisdom and advice to help you, the grant writer, make the world a better place.

Quotes:

Richard B. Ajluni, Director of Major Gifts, Catholic Charities of Santa Clara County

Robert T. Borawski, President, The Robert Brownlee Foundation

Elisa Callow, Planning Consultant and Former Arts Program Director, Foundation Program Officer, and board member

Emmett D. Carson, Ph.D., CEO and President, Silicon Valley Community Foundation

Mario P. Diaz, Community Affairs Manager, Wells Fargo Foundation, San Francisco Bay Region

Sandra W. Gresham, CFRE, Director of Philanthropy, Island Conservation

Meredith Dykstra Hilt, Executive Director, Tellabs Foundation

Diem Jones, Deputy Director, Arts Council Silicon Valley

Barb Larson, CEO, American Red Cross Silicon Valley

Honey Meir-Levi, Executive Director, Ronald McDonald House at Stanford

Veronica Patlan Murphy, Stewardship Director, San José State University

Marsha L. Semmel, Acting Director, Institute of Museum and Library Services

Grant Samples:

Rita Baum, board member and grants and appeals writer for Friends of the Library

Bay Area Glass Institute, www.bagi.org

Allan Berkowitz, Executive Director, Environmental Volunteers, www.EVols.org

Doug Cress, Vice-President Development, Orangutan Conservancy, www.Orangutan.com

Marilee Jennings, Executive Director, Children's Discovery Museum of San Jose, www.CDM.org

Beverly Lenihan, CFRE, *Reesults* Consulting

Belinda Lowe-Schmahl, Executive Director, Schmahl Science Workshops, Inc., www.schmahlscience.org

Beth Williams, CFRE, Director of Development, Hospice of the Valley, www.hospicevalley.org

There were also a couple of samples from experts who wished to remain anonymous.

Grant Writing 101

Part 1

Tips and Tactics for Generating Immediate Funds

1

Which Type of Grant Is Right for You?

Tin cup fundraising is antiquated. Modern fundraising gives donors the opportunity to invest in exciting initiatives and programs. Donors want a chance to change their corner of the world.

—Richard B. Ajluni, Director of Major Gifts, Catholic Charities of Santa Clara County

The first order of business for a new grant writer is to know what's out there. This chapter decodes some of the terminology that is used to describe the many ways grant dollars are awarded. Before we look at the types of grants, it makes sense to review the seven most common purposes for writing a grant proposal. Knowing your purpose will help you to narrow the search for the appropriate type of grant. Familiarity with these purposes also helps you to know what to ask for in your grant proposal.

Grant Purposes

Project Grants

If a grant guideline says that the organization only funds project grants, you want to make sure that's what you submit. This means that you ask for support of a specific project such as the study of penguins in the Antarctic, the acquisition of equipment for a computer lab, funding of a new art exhibition. A project has an end to it. These sample projects end when the penguin study concludes, the computer lab is equipped, and the art exhibit opens. Project grants are not ongoing. While the computer lab will have ongoing operating costs, your request for equipment is for a specific one-time expense. The purchase happens only once. While the art gallery will always have exhibits, bringing one specific exhibition to the gallery can be called a project. You bring it only once, even if the exhibit is there for a year.

Program Grants

Sometimes funders want to support ongoing efforts. This can be an education program at a zoo, a mentoring program at a YWCA, a job-preparation program with The Salvation Army, and so on. Programs are usually ongoing or multiyear. Funders often like to help implement new programs or expand programs when there's a need in the community. Program grants can help to supply materials, equipment, and brochures and pay for any number of events and activities. Some grants may cover staff costs. The thing to remember is that all the requested funds must go entirely to that specific program. Thus, if you want a grant for the salary of a program mentor and that person works in three programs, then only one-third of her salary can be requested in that grant or whatever portion you determine contributes to that program. As you'll learn in the coming chapters of this book, you ask only for the items identified as allowable expenses for the grant, whether they fund project or program grants.

Operating Grants

Every organization has operating costs, that is, the monthly expenses to keep the organization operational. This includes overhead such as rent, utilities, ongoing equipment rentals, and staff costs. The organization still has to justify the necessity of the costs and what impact the funds will have. The funder expects more than just keeping the organization going for a few months. Funders expect results, whether that is hiring a new grant manager to help increase income or paying the rental deposit for an organization to move to a larger facility to increase its reach in the community.

Organizational Effectiveness Grants

These types of grants help organizations to improve their way of doing business. You could request an organizational effectiveness grant to perform a study of your practices, to hire a consultant to help you develop a program, to carry out the recommendations of a consultant, or to implement your organization's internal strategies to improve your efficiency, competence, or outcomes in performing your mission. For example, say that your organization wants to increase revenue, and you're considering revamping your fee structure for the services your organization provides. You first may want to hire a consultant, who will bring expertise that your staff does not have, to guide you in your decisions. Such an expert will ensure that your organization doesn't price itself out of the marketplace, maximizes your income potential, and may even help with your marketing strategy to publicize the new fees.

Capacity-Building Grants

Funders want organizations to be sustainable, that is, to increase their capability to support themselves. What could your organization do to become self-supporting? That's what capacity-building funders want to hear about. Far too many organizations in the nonprofit sector don't look beyond the current year. They keep relying on the same income that they've always had. (But not guerilla grant writers.) You might request these funds to implement an annual campaign program, to purchase one of the corporate and foundation online directories that will help you to find new funders, or to build a new exhibit in your museum that will bring more visitors and thereby increase your revenue. Staff development (improving skills) could qualify as well.

Capital Grants

Capital expenses (anything above $100,000) are for things that have a long life, such as structures, buildings, and major equipment, although specialized vehicles also may fall into this category. Major renovations to buildings, such as adding an exhibit hall to an aquarium, count as a capital expense. The variety of opportunities means that there are a number of ways to request these funds. Some might require an extensive grant application process, whereas others may require a packet of your materials with a letter. Your research will reveal the way to make a capital grant request. Often the grant application is only the first step before a capital grant is awarded. Funders at this level are very committed to the organization and its mission or to the community the organization serves. They often will provide the funding in multiyear payments with checkpoints along the way to be sure that your project is on track. Some funders may require your organization to make the upfront payments and then reimburse you.

Endowment Grants

Endowment funds are meant to help sustain the organization or a specific program of the organization far into the future. The principal funds aren't spent; rather, only the interest from them is spent each year for general operating costs or earmarked purposes. Obviously, not every organization has endowments. Those that do may use a variety of approaches to grow the fund. They may write grant proposals, use annual appeals, make personal asks, and so on. Universities have alumni pledges. You would write a grant only after you've established a formal endowment program.

Now that you're acquainted with the various grant purposes, I can spring another concept on you. All the purposes just listed can have either a restricted or unrestricted constraint on them. *Restricted* means that there is a specific *use* within the purpose that the grant funds are intended for. In the preceding *project* grant example for the study of penguins in the Antarctic, a grantor may restrict the project grant to cover only the travel expenses for the participants in the study. Similarly, a restricted *program* grant for the job-preparation program may limit the funds to be used only on equipment purchases. On the other hand, an *unrestricted* grant to the Antarctic project means that the organization can use the funds for any costs related to that project, such as, travel, salaries, supplies, rental equipment, food, and so on. Whether the funds you receive are restricted or unrestricted depends on your proposal. What did you request? What did you convince the grantor you urgently needed the funds for? If you convinced the grantor that the project was vital to address a critical need, you may be awarded unrestricted funds. If you concentrated on describing one aspect of the project (that you determined suits the grantor's grant guidelines), you may be awarded restricted funds that can be spent only on that one aspect. It is considered something of an honor to receive unrestricted funds. This means that the funder truly trusts your organization.

Grant Types

There are several different types of grant proposals. The type of proposal you write depends on your needs, your time, available grantors, the grantor's requirements, when you need the funds, and your organization's policies. (Your organization may or may not have a policy that prohibits alliances with certain companies.)

A brief look at nine types of grants will give you some general guidelines as you begin your funder research. The lesson to learn here is to try as much as possible to match your resources (available staff time) to potential funding opportunities that are a natural fit and lead to a favorable probability of getting funded.

Corporate Grants

Many major corporations have a corporate giving program that has formal grant procedures with applications and deadlines. Corporations identify their current funding priorities in their guidelines. They may reveal that they do not fund capital grants or that they only want to see program grants. They may state that they focus on education projects or, more specifically, science education. In most cases, a grant committee of employees makes funding decisions. Sometimes the grant function is managed by a corporate foundation that still will follow the directives of the corporation. Most corporations give the bulk of their support to the local areas where they have employees. Some give to regional or national causes. So a corporation such as Walmart that has numerous stores throughout the country is more likely to give in all those areas where the stores are located, whereas a corporation such as Dell, Inc., that has a presence in six states, only accepts grant applications from the counties where its offices are located. Corporations also may give to organizations or causes that have a tie-in to the product(s) they produce. For example, Microsoft Corporation gives globally to fund technology learning centers and technology job training, projects that directly match its business. A little preliminary research goes a long way to avoid wasting time on an application for a grant for which your organization doesn't qualify.

Points to keep in mind with corporate giving are that there are usually formal applications with a definite deadline date and a definite award date. Grantors may have focused priorities for their grant dollars and often spell out the specific types of requests they will and will not consider within that focus. They likely have specific award limits. Therefore, looking at the factors mentioned earlier, if you need the funds right now rather than six months from now, your program is not listed as a high priority for the grantor, and you need more than the $10,000 maximum allowed by the grantor, then look elsewhere for your funding for this particular grant need.

Foundation Grants

Foundations have more flexibility than corporations. Since corporate foundations often still have the same restrictions or requirements as mentioned above, this paragraph will discuss the noncorporate foundation. Yes, private foundations do have formal procedures and maybe even an application process, but they are more apt to move quickly if necessary. Foundations have board members who make the funding decisions. While foundations will have their own set of guiding principles and funding priorities, they often allow a lot of leeway in the proposals they will consider as long as the request falls within those guiding principles. Discretionary funds allow a foundation to fund an exciting project that may not meet its funding priorities but presents an outstanding opportunity for the foundation to make a difference. Sometimes the board members themselves will rally for a particular organization or a project grant proposal that has been submitted. Foundation grants vary in size from the very small to the very large, from $1,000 to hundreds of thousands or even millions of dollars. There are countless family foundations, and they may not have Web sites to allow you to find basic information. In such a case, a call will be necessary to get vital questions answered before you write the proposal. Depending on the foundation, you may have to submit an application or simply a two-page letter. Proposals may be accepted year round at one foundation and have only one specific submittal date at another foundation. If your proposal fits the principles and mission of a particular foundation, then foundation funding may be a good fit for your grant need.

Government Grants

Municipal, state, and federal governments offer a variety of grant opportunities. As one might expect, these applications are often rigorous and time consuming to complete. Seriously weigh the time investment against the potential gain if your grant is selected. No one has reporting requirements like the government. On the other hand, millions of dollars in grants are awarded by government entities every year, such as the Department of Education, the Department of Homeland Security, the National Endowment for the Arts, and the National Science Foundation, and they may fund projects no one else will, for example, new restroom facilities. Government entities have their own grant priorities, such as access for disabled people, which may mean that support is available for your project.

Individual Grants from a Family Fund

The way to get a check from an individual is to ask. If you don't ask, whether in a letter or in person, you won't receive. When you do ask (as in a mail appeal), though, and a check arrives with the name of a family fund or family trust, what does it mean? It means that one of the individuals within that fund or trust received your request. The group of people who have pooled their funds into this account likely have voted on and approved the funding. There isn't an application process for family funds. You request from and cultivate these funders the way you would individuals. Asking in person is the subject of many books written by other authors. Asking for support in a letter should be brief to reflect the available time of busy individuals. Try to state your case for their support in one to two pages following the examples of short requests in Chapter 14. Use this type of request when you know that an individual has an affinity for your organization and has the ability to fulfill the request. You also use this method when someone within your organization or board knows the individual personally.

Donor-Advised Funds

Some donors create donor-advised funds to assist them with their philanthropic giving. This type of giving can be considered one step short of the donors creating their own foundation. These funds are held by a community foundation or brokerage that lets the donor recommend where their charitable dollars go. Donor-advised funds are less time intensive for the donors because they are not involved in the day-to-day management of the funds or the reporting requirements to the Internal Revenue Service (IRS). Donors leave the professional investment management to the community foundation or brokerage to which they entrusted their funds. Donor-advised funds vary greatly in how they disperse grants from one fund to another and in how they operate from one organization to another. It is best to view the Web site of the organization managing the funds to learn how to request grant funding. These organizations have officers who can answer your questions. While some donor-advised funds do not allow grant applications (they fund only the organizations selected by the donors), others allow grant application requests. In such a case, you would follow those grant guidelines. Sometimes there's a call to the community, announcing the availability of money for a specific cause. This likely came from a donor's stipulation of how their funds should be allocated. As a starter, to find donor-advised fund opportunities, do an Internet search on "community foundation + *your city*." And also try, "donor advised funds + *your city*," for example, "community foundation Seattle" or "community foundation Detroit." Then do a search on "donor advised fund Seattle" or "donor advised fund Detroit."

Charitable Trusts

There are different kinds of charitable trusts that individuals can set up as part of their estate planning. At the time of the person's death, an administrator handles the financial wishes of the deceased as predetermined in his or her will. You only need to know that sometimes those wishes are for grants to be awarded in the community. As with other grants, sometimes there are restrictions as to which types of organizations or causes the grant awards will support. Some charitable trusts can be quite large, such as the J. Paul Getty Trust. Read the guidelines, and if there's a match, apply.

Sponsorships

Sponsorships are a bit different from outright grants in that the sponsor receives benefits for the gift. The gift can come in many forms, depending on what it is you requested. You can ask for a cash grant, of course, but you also can seek a media sponsor, a program sponsor, an event sponsor, and so on. Sponsorships are great for creating win-win situations that help you to cultivate a positive relationship with the sponsor. Consider sponsorships when you have a specific need for a specific project, program, or event and you have identified benefits to the sponsor that make sense for the level of grant you are requesting. For example, if you want a T-shirt sponsor for your group's basketball league, and the cost of purchase, printing, tax, and delivery is $600, what can you offer the sponsor that is about $600 in value? Can you offer a full-page ad in your season's brochure or schedule? Season tickets? An acknowledgment over the loudspeaker at the season opener? A banner (that the sponsor supplies) displayed at all home games? Get creative in what you seek sponsorships for and what you can offer a sponsor. Potential sponsors are everywhere. They are local businesses; service organizations such as Kiwanis International, Rotary International, and League of Women Voters; chain stores such as Target Corporation; local utility companies; local government agencies; and any size corporation. Even small, one-person, home-based businesses could use local publicity. Connecting sponsors to a project, program, or event that matches their business purpose or philosophy in a creative way and offers benefits to the sponsor to boot, creates an opportunity that's difficult to pass up. Write a persuasive letter that is no more than one or two pages. As with all short letters or grant proposals, you can always add a sentence that says that more information is available on request, or you can include attachments.

With sponsorships, try not to offer everything up front. Leave something for negotiation or for a surprise. It's great when you can give your sponsors (or grantors) a little something more than they were expecting. In the basketball league example earlier, if the sponsor agreed to a $600 level sponsorship for the banner, acknowledgment at the season opener, and a half-page ad in the season brochure, you could surprise the sponsor by upgrading the ad to a full page or adding the sponsor's business name to the team T-shirts. The sponsor would be pleasantly surprised!

In-Kind Grants

In-kind refers to noncash grants. This type of donation can come in a multitude of forms, including donated equipment, supplies, food and beverages for an event, media advertising, vehicles, and anything else in a proposal letter that you can convince someone to donate. Services such as Web design, printing, landscaping, and painting fall under this category too. Let's lump those requests for admission tickets, gift baskets, weekend getaways, and similar items for your auction in this group. Following the criteria of sponsorships, if you can entice businesses, corporations, or individuals to donate in return for benefits you've preestablished, however small, then send them a brief in-kind grant request. Follow the short grant format found in Chapter 14.

Matching Grants

A matching grant can be combined with any of the preceding purposes and types. As the name implies, the grantor is matching the funds that come from another source. That source can be another grantor, your organization's own funds, capital fundraising, or an appeal to the community. The grantor determines how the grant funds will be matched. Some may provide their matching payment after your organization raises the agreed-on amount. These opportunities often have deadlines by which the funds must be raised in order to receive the match. Only apply for a matching grant if your organization has the ability to come up with the matching funds and meet all the conditions of the match.

Throughout the remainder of this book I'll use the term *grant proposal*, and the same concepts apply regardless of which type of grant you're considering.

Summary

Guerilla grant writers apply for the types of grants that suit our particular identified needs, taking into account our time, available grantors for the type and purpose, the grantor's requirements, when we need the funds, and our organizations' policies. We apply for grants from organizations that are a suitable fit with our organizations.

2

Understanding the Components

Grant Writing 101 is a very basic presentation of: Who, What, *Why, and When. Regardless of a funder's format, you must clearly articulate:* WHO *you are with a bit of historical reference,* WHAT *you do as a mission or statement of purpose,* WHY *you are responding to the needs of your target audience, and* WHEN *you deliver your products. This basic formula works for individual artists, as well as arts organizations. After you create your draft, make sure you get an outside reader to review and comment* OR *read the entire narrative out loud.*

—Diem Jones, Deputy Director,
Arts Council Silicon Valley

The purpose of this chapter is to prepare you for the types of questions you might encounter in proposal writing. Even if a specific question isn't asked, you may want to address the topic in your letter or proposal.

In any case, these items give you something to think about and inform you of the types of data your organization may want to start recording. A looming grant deadline is not the best time to ask staff to gather and send their information to you. Nor is it the time to try to create a mission statement. No one wants to pass up an incredible opportunity for funding, so work on these items before they're needed. Skillful planning leads to guerilla grant writing.

No one grant application will ask all these questions. But they will ask some or a variation of these questions, so keep this book nearby for a quick reference. You can be as brief as a few sentences or as long as a couple of paragraphs. Or you may need several pages for a particular question. The length depends on how much detail is needed to provide valuable information to the grantor. Often your length is predetermined by the grant application. In general, brevity is best.

Organization Questions

Overview of Organization

This is where you give a description of what your organization does and for whom. You may include your mission and a very brief history of your organization if these are not asked for in separate questions. If there is any significant news to report (e.g., you just dedicated a new pediatric wing, your dance troupe just completed a European tour, or you just served your ten-thousandth meal), add that here. This section introduces your organization to the funder. It tells the funder interesting facts about who you are, how long you've been around, and what makes you special.

Organization's History

Even if you work for an organization that's been in existence for 50 years, you can sum up the history in two paragraphs. Highlight the turning points. Include key accomplishments. You might want to make comparisons if you have them; that is, "In 20 years we went from 400 clients to 30,000" or "Since we opened our doors in 2006, we have served 900 seniors with no-cost physical therapy treatments." Whatever you write should make known your organization's value to the community where you are located.

Your Mission

If your organization doesn't have a mission statement, it needs one. Your mission expresses the reason(s) you exist. (Many of the Web site resources identified in Appendix C offer online help in creating one.) Your statement hopefully is no longer than a couple of sentences. If the application allows, add a sentence or two more to state your objectives or vision. Perhaps you want to include the program's mission as well. Even if you are not asked for your mission, state it clearly somewhere in your proposal. Your mission is fundamental, and grantors expect you to know that.

How Does Your Mission Align with the Funder's Core Priorities?

As discussed in Chapter 3, research the funder before you begin writing the proposal. Knowing the funder's priorities helps you to write a stronger proposal. By addressing how your project ties to the funder's mission, core principles, or funding priorities, your proposal moves one step closer to consideration for support.

Example

"Our proposal is compatible with your grant making priorities to support basic needs of food and shelter. With XYZ's grant support, together we can ensure our common goal, that women and their children have a safe refuge, by making possible emergency housing for domestic violence victims."

Primary Focus of Organization

Many organizations provide several services. An aquarium, for example, may have its living exhibits, a conservation program, and a research function. A hospital could have an acute-care medical facility, an accredited nursing college, and a rehabilitation center. Name your primary focus areas or key programs. If this question were not asked specifically, you'd still want to identify your focus areas in your overview or history.

List Your Organization's Accomplishments

Highlight your accomplishments, including any awards and special recognition. You may highlight milestones of your organization, such as "In 2010, we swore in our one-millionth scout," "We saved the life of our ten-thousandth prematurely born infant," "In 2010, we celebrated our sixtieth anniversary," and so on. Did your visitorship grow from 250,000 to 400,000 in five years? Was your organization featured on the front page of a major newspaper or on a TV show? This is the section in which to toot your organization's horn.

Describe the Benefits Your Organization Provides to the Community

If this question isn't asked, you need to address it. You likely provide many benefits to the community beyond this one project. Discuss how the services you provide benefit others, the gap you fill, and the ways you improve the quality of life of the residents in your community. If you're applying to a corporation, don't forget to mention that you serve its employees (if you do) or the communities where the corporation's employees live. If your location is near the corporation, then it's safe to assume their employees live nearby.

How Do You Collaborate with Others?

Funders approve of collaborations to serve the community. They show that you are not duplicating services, that the problem is important enough to incite another organization besides your own to participate, and that you have partners with whom to combine resources to address the need. Partnerships give you additional credibility. To answer this question, state the names of the collaborating organizations and what their roles are. Think about the collaborations you may already have but to which you haven't applied such a description. Do you provide services to another nonprofit at no charge? Does another nonprofit provide services to your organization at no charge? Do you join forces to stage an event? Do you partner with schools? Does a media outlet give you free publicity for this program? Perhaps a local business provides the supplies you need for the project. A variation of this question may ask you to identify other organizations involved in your project.

Example

"We collaborate with elementary schools in the community to provide enriching afterschool arts opportunities to underprivileged children. The organizations include. . . ."

List the Members of Your Board of Directors and Their Affiliations

Name each board member, and include his or her employer's name and position/title. Also identify his or her role on your board. Are they members at large or officers? Some applications ask for contact information on board members, so an updated board roster in an electronic form is handy to have. Some funders ask for each board member's length of service on the board. Keep this roster updated so that you're not scrambling when you need it. Include your organization's executive director's contact information in this response if that person isn't signing the cover letter or included elsewhere in the application.

What Are the Strengths of Your Board of Directors?

This is a good question and one with which you may need assistance. Jot down your initial thoughts, and then ask the question of the board president, executive director, and your manager. See if they have things to add or if they can enhance what you've stated. This is not an easy question to answer if you don't work closely with the board or hear about their accomplishments and activities on a regular basis. Perhaps their strength is their fundraising prowess, or an annual event they coordinate to raise funds for and awareness of your organization, or their commitment to serve the neediest families in your community by sponsoring their fees to receive your services? Is their strength their role as ambassadors for your organization seeking community partners that result in sponsorships and partnerships? Do they increase the visibility of your organization by their activities, such as participating in community events or accepting speaking engagements?

Example

"Our board members bring a wealth of public and private-sector experiences and community connections to the board. Our directors meet with prospective donors at site visits, and they cultivate current supporters, resulting in the acquisition of donors and thousands of dollars being raised from both groups."

Example

"Our board provides operational funding for events, programs, and small capital improvements. Their dedication, since 1985, has been invaluable in providing grant funds, additional visibility in the community, and ambassadors for our museum."

Qualifications of Persons Administering the Project

Another way of asking this might be: Who are your key staff for this project? The funder wants to be assured that your organization has the appropriate personnel to execute the project as well as manage the project. Briefly describe the relevant experience and credentials of the key staff related to the project. Some applications may ask for their résumés. See the following example that one aquarium might submit of three staff members.

Example

Aquarium Director of Education—Barbara Smith

"With degrees in oceanography and education and experience in aquarium management, Barbara came to ABC Aquarium in September 2000. In her 20 years of experience in informal science education and curriculum development, Barbara managed two reaccreditations of the aquarium with the Association of Zoos and Aquariums; the development, delivery, and evaluation of education programs reaching over 2 million aquarium visitors; and the development of the aquarium's conservation research program. Barbara will be responsible for the overall direction of the project."

Aquarium Education Program Coordinator—Bob Smith

"Bob joined the staff at ABC Aquarium in 2006 with five years of experience in environmental education and children's camp programs. A graduate of UC Santa Cruz in environmental studies, Bob has worked as an education specialist for both EFG Aquarium and HIJ Aquarium, focusing on aquatic wildlife conservation education and curriculum development. Since Bob arrived at the aquarium, the variety of educational programs and the number of participants has doubled. Bob is dedicated to the enrichment of children's learning about marine science and leads fun activities such as our whale-watching tours and oceanography camps. Bob will supervise the development of educational activities, curriculum development, and instructor and docent training."

Aquarium Education Program Instructor—Joan Smith

"With a bachelor of science degree in marine biology, Joan is an enthusiastic teacher of marine science to visitors of all ages. She completed a summer internship at EFG Aquarium before her arrival at the aquarium in 2008. Joan teaches over 200 school groups about marine science and conservation every year. She helps to develop special discovery labs that enhance the ocean learning students receive in school. Joan will implement and lead the project with docent assistance."

What Are Your Organization's Strategic Goals?

Does your organization have a strategic plan or long-range plan? It may be called the "five-year plan" or something similar. Basically, these are long-term plans beyond one year. Include a paragraph or two that sums up relevant points. Also state any actions you've taken to accomplish these goals. Are you halfway there?

Describe Recognition Opportunities

What plan do you have to recognize your donors? This can be as simple as adding their names to your newsletter and Web site or adding their names to your onsite donor wall. Recognition is different from acknowledgment. The latter is the thank-you letter you send immediately on receipt of the donor's grant check. Recognition is the public thank you, and ideally, it should correspond to the size of the grant. Realizing that your organization may have limits on how and where you can recognize, you'll have to find creative ways to do so. One organization that lacked a donor board printed a beautiful poster that listed all their donors' names and hung it inside a large display kiosk at their facility's entrance. There are numerous, fitting ways to thank your donors. Whatever ways you decide are appropriate for this grant (and eventually all others), inform the grantors in your response.

Example

"We will be pleased to add XYZ's name to our donor board for a year. XYZ also will be recognized in an upcoming issue of our newsletter and on our Web site."

Note: A variation of this question is for the grantor to ask about the visibility plan for the project. While you can include the visibility you'll give the donor, they are interested in the visibility of the project so that the primary clients you intend to serve will know about it and can come to you. A grant does no good if the intended users don't know about the services.

How Do You Serve Underserved Populations?

Describe how your program or organization serves underserved people. You may serve these populations directly as part of your business. Or perhaps you serve them indirectly with discount admission days, or provide free programs, or offer services in neighborhoods to reach the underserved. Or you're located conveniently next to public transit or offer free parking. Do you come to them, perhaps working in collaboration with a community center or church to reach the underserved who frequent those places? Think about how underserved people access your services.

Who Are the Target Populations You Serve?

Describe your primary and largest clients, visitors, or users. Include whatever demographic information you have, such as the age groups, income levels, ethnic backgrounds, residential areas, and any other qualities of the people you serve. You might find that your local government, such as the city or county, has statistics that you can use. The city or county Web site may contain demographic data that pertain to your project, such as income levels and ethnic makeup of the city or county.

Example

"The primary audience for our project is preschool-age children from non-English-speaking families. A recent study conducted by the county revealed that this target audience is expected to grow 23 percent by the year 2020. This study also indicated that these families' household incomes were below the county average."

Discuss Volunteer Involvement

Committed volunteers are another sign of an organization that the community values. Mention the number of volunteers or the number of volunteer hours and the makeup of the volunteers. How long has your volunteer program been in existence? Do you have a formal training program for volunteers? Describe any other distinguishing facts that may interest a funder.

Example

"We are pleased to have a volunteer force of 30 parents who provide over 3,600 hours of classroom aide service each year."

Example

"Our volunteer program provides job experience and mentoring opportunities for 20 teens each year. During the nine years it has existed, it has given 175 teens invaluable leadership and socialization skills, as well as practical administrative expertise."

Are There Volunteer Opportunities?

Funders ask this question only if they have an employee community volunteer program, which shows their genuine interest in volunteer opportunities with the organizations they support. Think of ways you can use volunteers for either short or long-term projects. Would you like more corporate presence on your board of directors? Could you use a team of 20 to complete a paint project in one afternoon? Brainstorm how these employees can help you to fulfill your mission. Give them something meaningful to do.

Example

"XYZ Corporation employees can get involved with ABC Aquarium by participating in our conservation programs. For example, employees can organize an ocean cleanup day in conjunction with our cleanup event. Or they can coordinate a cell phone drive—by collecting unwanted cell phones from XYZ Corporation employees. Both efforts would make a positive impact on the environment. We also host an ocean appreciation event in June that requires over 100 volunteers. We would greatly appreciate XYZ Corporation employees' help and participation in preserving our oceans."

Example

"Volunteer support is, and always has been, a significant factor throughout ABC Zoo's 65-year existence. Our education program involves both docent (adult) volunteers and teen volunteers.

> The docents are an integral part of our outreach program to schools and organizations in the community. Volunteers also provide onsite educational demonstrations to guests in the zoo and assist in onsite classes and camps. Our volunteers who complete the required training and work in the education program enjoy working with children and animals and find their experiences at ABC Zoo rewarding. In addition, we are seeking a volunteer from XYZ Corporation to fill a two-year term on our board of directors. Please contact our executive director for more information about these exciting opportunities."

Project Questions

Provide a Single-Line Description of the Project

This may seem like a simple question. But you want that one-liner to grab the funder's attention. It should sound compelling or exciting. If your request is funded, this one line may be the only description the funder uses on its Web site or in its annual report or employee newsletter. You want all those audiences to see your organization's name beside that description and think, "*Wow!*" On the other hand, don't lose sleep over it. Get the proposal written, and come back to this. Something may come to mind during the course of creating the application. For hints, look at your mission statement and your project description paragraphs. If necessary, send the proposal out with your plain description.

Example

"Expansion of the Science Project Program to provide enriching veterinary science opportunities to primary school students."

Example

"Develop a disaster preparedness plan and train the ABC Nature Center's staff, volunteers, and board of directors in disaster preparedness and response.

Proposal Title

The same thing said about your single-line description applies here. Your title may be the only description beside your organization's name. A potential funder should be able to look at your title and/or one-liner and know that your proposal fits its organization.

Provide a Summary of the Project

You get a few lines to summarize the project. Make those sentences work! Give a brief overview of the proposal. You can stress the need along with your solution (your grant request is the solution) in about three to four sentences, right? Then you use the rest of the proposal to flesh out these points.

Example

"On behalf of ABC Art Museum, we respectfully request $50,000 for the renovation of the exhibit hall. This proposal will create our first children's area, where children can express their creative side through artistic expression. Children will learn through a range of activities, including painting on a paint wall, creating sculptures, and playing music. These activities will immerse children in art and allow them to use their imagination, motor skills, and reasoning skills."

Describe the Project

This is the meat of the proposal. It's where you communicate what you are going to do with the funds. Emphasize the purpose of the grant. It's important to express the substantial need for your project in the community you serve. Describe whom and how many the project will serve—and for how long. It's essential that you clearly put into words how this grantor's funding will have a significant impact. Be concrete. No vague statements. If you served 3,000 meals last year, has the need grown this year? Will the funder's grant mean that you don't have to turn away 150 hungry clients per month? If the following questions are not asked separately, you may want to address one or more here:

- What have you done already to address the need?
- What are the project's goals and expected outcomes?
- How does your project align with your mission?

- How does your project align with the funder's core priorities?
- Who are your key staff for this project?

Examples

Vague

"Your support of this project will greatly improve the lives of 100 troubled teens."

Concrete

"Your support will provide six months of individual counseling, GED tutoring, and job training for 100 teen dropouts."

Okay

"Though we serve 3,000 meals a month, the need has increased."

Better

"Though we serve 3,000 meals a month, we are forced to turn away 90 families each month due to the increased demand for our services."

Best

"The expected result of this project is to serve the most disadvantaged families from our community. Your investment of $15,000 will provide 9,000 meals, helping us to meet the growing demand for our services. With your support, we won't be forced to turn away the 90 families each month that we currently don't have the resources to feed."

Statement of Need, or Describe How Your Project Fills a Need

You included the need in the question that asked you to describe your project. An application won't usually ask both questions, but if one does, it gives you another opportunity to offer convincing evidence that your proposed project fills an urgent need. Convey how your organization is the hope for the community. Another way to pose this question may be to ask you to summarize the issues addressed by the program. Remember, the need you're filling is not your organization's need, but rather a need in your community. In the following examples, one describes a need for computers for the agency's lab, and the other describes a need for computers to serve the community that also addresses the issue of unemployment. Which one sounds more urgent?

Examples

Weak

"We urgently need funds to buy computers for our lab because the ones we have are always in use, and clients have to wait over an hour to use one."

Strong

"Through our computer lab, we provide resources for our unemployed clients to find work. With county unemployment rates the highest in five years, the demand for our services has increased by 28 percent over the past year. Additional computers will allow clients to perform a thorough job search, write and print résumés, and apply for jobs online."

How Does Your Project Align with the 40 Developmental Assets?

Some institutions are more likely to come across this question than others. And the "developmental assets" apply only to those serving children between the ages of 3 and 18 years. If you've never heard this term before, it's time for a little Internet research. According to Search Institute, "the Developmental Assets represent the relationships, opportunities, and personal qualities that young people need to avoid risks and to thrive." Search Institute provides the lists it has developed for four different age groups at www.search-institute.org/developmental-assets/lists.

With a little concentration, you'll find ways to describe how your organization's activities or programs align. Here's the thing: once you've identified activities that your organization performs that fit one of these assets, write them down. You'll end up with a power sentence. Do this for several of your activities, and you'll have descriptions that sing.

Example

"The ABC Homework Center provides a caring, encouraging environment for middle school students to study, receive tutoring, and learn."

Example

"The Teen Achievement Project increases the social competencies and self-esteem of 30 eighth graders by providing opportunities for them to learn conflict resolution and decision-making skills with adult role models."

Discuss the Program's Background

This section allows you to describe the program for which you are seeking funding, that is, the program under which your project falls. Your description should illustrate the successes and challenges, which relate to the reason you are requesting support. You can discuss your program's growth. Give a concise overview of the program. This section convinces the funder that your organization can pull off the project. If your program is new, you might discuss why you started the program—in response to what need?

How Will the Funds Be Used?

As mentioned in Chapter 5 and throughout this chapter, personalize your numbers as much as possible.

Example

Not

"Your support will provide 100 cases of formula."

But

"Your support will provide formula to feed 100 babies for one month, serving low-income families from our community, and having positive effects on the wellbeing and early development of their young children."

Funders want to see what their grant dollars are funding. If the funds will pay for more than cases of formula, state that as well. Keep in mind allowable expenses.

Example

"In addition to 100 cases of formula to feed 100 babies for one month, funds will provide supplies such as reusable bottles and bottle cleaners."

Example

"In addition to increasing the homework center hours, XYZ funds will provide two new desktop computers and two algebra tutorial software licenses."

If the budget is short like the following example, you may include it in the body of your proposal.

Example

ABC Homework Center Project

Increase hours to 234 hours	$3,600
Two computers	$1,900
Algebra software licenses	$500
Total request	$6,000

Note: The reason guerilla grant writers make calls to prospective funders is to ask about their proposed projects. Through your research, you may find that the funder does not directly pay salaries. Yet you see that the funder did just that with previous grants. The way those proposals may have been presented was to include the salaries in the "project." Notice that I didn't mention the salaries of the homework center tutors earlier; rather, I presented the homework center *project,* which hinged on increasing the homework center's hours. Call the grantor's representative *before* you write the grant proposal to determine if this tactic is allowable.

Duration of Funding Impact?

Funders want to know how long their funds will last. Will their grant cover the intended project for three months, six months, or one year? Your proposal should include the duration of funding if you were to receive the proposed amount. Most funders will not provide a grant for more than a one-year duration. Their guidelines will inform you of the funder's limitations. If your project truly is a two-year project, it is best to talk to the program officer to see if two-year funding is allowed. If not, do not request more than one year. Ask if there is potential for renewed funding. Or can you reapply for the second year? After your discussion, you can decide to either continue looking elsewhere or apply for one year of support.

Note: You may receive a smaller amount than you requested. If so, you obviously will need to adjust some aspect of your project. Either the duration will change, or the number of people you serve or something else will change. You do *not* need to write those options in the proposal.

Number of People Served

Another essential element in your proposal is the number of people your project will serve. While you want to plainly state the number of people and duration, as noted in the following example, you'll also want to elaborate on the people served when providing the project description. Tell us something about them that relates to the need you're filling.

Example

"Your generosity will permit 90 high school students to enhance their musical abilities and participate in a school band, providing a year-long opportunity to build teamwork, get exercise, and increase their self-confidence."

What Is the Geographic Reach of Your Project?

Do you serve one neighborhood or the whole city? Maybe your project reaches the entire county or more than one county. This is a straightforward question. However, if you will offer elements of the project online, then your reach has increased dramatically. If you will share your project's lessons learned with other like institutions, or if these other institutions will be able to emulate your project, again, your reach has increased.

Timeline for Implementation and Completion

While this question is really useful when the request is for brick-and-mortar capital projects, it also applies to other projects, particularly new programs.

The idea to keep in mind is to ensure that your completion date falls within the funder's grant completion period or you'll knock your proposal out of the running for support. Make sure that the duration matches the responses you provided in other parts of your proposal.

Example

"The project will be initiated within 30 days of the receipt of the grant award and will be completed nine months from that start date."

Discuss Your Evaluation Plan

A variation on this question is: How will success be measured? Your organization may already have systems in place that track the numbers. These could include your total visitors, the number of kids enrolled in camps, the pounds of food distributed, or the quantity of meals served. Your evaluation includes what difference the grant will make. Before a funder awards your organization a grant, they will want to know how you will know if the funds have succeeded and have had the desired impact. The evaluation is quantitative. The funder wants to see numbers that mean something. Tell them specifically what data you'll collect or criteria you'll use to evaluate the project. Along with this, say that you'll provide the funder with a final report with the actual outcomes at the end of the project. Then do it. However, if your organization intends to hire an outside evaluator, be sure to include that choice here as well as including the cost in the budget.

Example

"With your support of this project, we can increase the number of outreach visits to schools by 20 percent, serving 3,900 children in one year with enriching art workshops. Evaluations by the classroom teachers will rate the workshops to measure the increase in the children's knowledge and appreciation of art. Specifically, the teachers will evaluate three areas: (1) At least 80 percent of children can identify the art forms presented, (2) at least 90 percent of the children participated in creating their own art piece, and (3) the art instructor received at least an 85 percent rating for program quality and effectiveness."

Example

"With your funding support, we can increase the number of homework center hours by 25 percent, allowing us to provide 234 hours of homework help to XYZ middle school students for an academic school year. [In your project description, you will have identified the number of students the center helps each hour.] Our evaluation will focus on two areas: first, the number of total students served and the number served during the added hours, and second, student evaluations of the facility and tutors. We will review the feedback from the evaluations and make any necessary adjustments. Our goal is to reach 3,510 students and to achieve 90 percent positive student evaluations."

Note: While you want to evaluate the project's effectiveness, try not to create an unnecessary burden on yourself or the staff involved. Perhaps there's a way to use data you already collect? In every case, make sure that the evaluation method is not cumbersome and that the information is useful to your organization and your funder.

Financial Questions

Who Are Your Funders? For Organization? For Project?

We have discussed partnerships several times in this book. A funder wants partnerships, too, and often establishes relationships with nonprofit organizations (who are a match) because of the relationships they have. Not including individuals, list all the funders or sponsors your organization has. Some applications may ask based on the project, some for the organization as a whole. This question also lets potential funders know that someone else in the community has supported you in the past, which implies that you're aware of meeting grant requirements. And grant support from others also affirms that someone else found that your work is valuable to the community.

Who Else Have You Asked to Support This Project?

List any outstanding requests related to the project. List the organization's name, the project's name or description, and the amount. It's possible that you are requesting support for only one element of the total project. You would add these requests, too. For example, say you submitted a proposal for $30,000 for a library van. Then, in answering this question, you might identify that you submitted three other proposals for support of this library van project. For example:

1. ABC Foundation	Shelving and lift for library van	$10,000, pending
2. EFG Foundation	Books to stock van	$ 2,500, pending
3. HIJ Corporation	Library outreach support	$ 5,000, pending

If you have received any funding for this project, you would definitely add those names, too. Identify any funding that has been committed or received and the amount.

Note: Here's a tip about grant award limits. Your selected funders may only offer grants up to $5,000, but what happens if you need $10,000? It is entirely acceptable to ask two funders for $5,000 each for the same project. You would make a note of the dual request in your grant proposal. A potential funder also needs to know if you can conduct the project if you do not receive the other grant. So a sentence or two addressing this concern will strengthen your proposal.

In the preceding budget sample, the grant writer strategically asked for three grants based on her research. ABC funds equipment, EFG provides only project grants, whereas HIJ provides general operating support and project support. Sure, it's preferred to ask one funder for the total amount, but that is not always possible.

What Will You Do If You Do Not Receive Full Funding?

Funders receive many more requests than they can fulfill. One way they can support as many projects as possible is to make partial grant awards. Give this question serious thought. Do not state that you can proceed with less support unless you truly can. This is linked to never promising what you can't deliver. Likewise, think before saying that you cannot offer this program or provide this service without full funding. Instead, consider an adaptable response. Perhaps you will serve fewer clients if the award is lower than the request. Can you remove one element of the program without a severe impact on the quality or intention? It's also suitable to state that you will continue to seek support from the community for this project. In other words, you will write more grant proposals or attempt other fundraising approaches to achieve the full funding needed.

Examples

"If the project is not fully funded, we will continue to seek support from the community through grant proposal writing."

"If the project is not fully funded, we will continue to seek support from the community through our annual telethon fundraiser."

"If the project is not fully funded, we will provide the literacy program but will not offer the free book bag with books to our clients unless funds can be secured from other sources."

Describe Your Sources of Funds

Your sources of funds may come in a variety of ways or just one or two. If you have several sources, you may wish to use a table or a simple list as in the following example.

Example

Percentage of Total Sources of Funds by Contributor Type

Contributor Type	Percentage
1. Individual donors	9%
2. Corporate cash grants	35%
3. Corporate in-kind contributions	1%

4. Foundation funders 40%
5. Government sources of support 3%
6. Other* funding sources 12%

*Other funding sources includes income from memberships (about $84,000 per year over last three years), annual operating income from event ticket sales (about $37,500 average over the last three years), and interest (about $2,617 average over the last three years).

Attach Your Organization's Budget

Naturally, potential funders want assurances that your organization will stay afloat long enough not only to complete the project but also to continue your work for years to come. When they award a grant, they are investing in your organization. They want that investment to have long-lasting effect, even after the funds have been spent. Your budgets reveal your size, your priorities, and your funding sources, among other things. Your budget documents are another of those items that should be prepared and ready to submit with applications that request them. An application may ask for your budget in one, two, or all three of the following ways: organization, program, and project.

Example Using a "Friends of the Library" Organization

Organization budget—This includes the entire nonprofit organization's budget.

Program budget—This would single out one program from the many that the organization provides, such as an author speaker series, an adult literacy program, or a children's reading program. Select the program related to your grant proposal.

Project budget—This budget is specifically for the grant project you're writing about. In this example, it may be a grant of $2,500 for a Harry Potter Summer Fun Project (that falls under the children's reading program), and you would show how the $2,500 will be used to ignite the imaginations of children while increasing their joy of reading.

How Will You Sustain the Project?

Project sustainability is as important as organization sustainability. Describe how the project will be funded in the future. Even if you offer a program free of charge to the public, there must be a way for your organization to pay the rent and salaries and other costs to keep the project going. Are some costs paid by a government multiyear grant? Do you have memberships that are renewed annually? Does your organization conduct an annual appeal to the community? Do you charge for some services that raise the funds to run all your other services? If you intend to write grant proposals or use other fundraising methods to sustain the project, that should be mentioned as well.

Example

"The project will be sustained through operating revenue, including admission fees to the summer orchestra series, sales of series CDs, advertising in the series program, and by community support through grant proposal writing."

Describe Your Financial Stability

Does your organization have an endowment? Does the organization have a dedicated cash reserve? These are special funds that help to ensure the longevity of a specific program or your organization, and they show that your organization has planned for the future. But instead you may use other means that speak to your financial strength. Perhaps you have strong community support in the form of an annual appeal or membership program that has increased steadily over the past five years. Community support is crucial. How does the community support your organization financially? Through workplace giving campaigns? Perhaps you have the support of your state government in the form of a 10-year grant for operations. Just as impressive is a statement that you have managed your funds well over the last eight years (or since your existence) and have balanced your budget each year. Also mention whether you do or do not have accumulated debt. Elaborate on any potentially negative matters regarding your financial history and what your organization has done to rectify the situation. You want to gain the funder's confidence in your organization's financial viability.

Example

"ABC Aquarium has been in existence since 1980. The Aquarium Foundation provides funding for operations and maintenance. The majority of funding comes from operating income (admission fees, memberships, and formal education classes and camps). Other income comes from grants from foundations and corporations and donations from individuals and organizations. ABC Aquarium has cash reserves. We do not have an endowment, nor do we have accumulated debt."

Administrative Questions

Contact Information

Once in a while, you'll come across an application that asks a series of questions but doesn't ask for contact information from the applicant. In such cases, include a cover letter with the name, e-mail address, telephone number, and Web site address of your organization if these are not requested in the application. It's a good idea to give an alternate telephone number. Provide either the main organization number or the executive director's number in addition to your own.

Cover Letters

A cover letter is nice to include, if they're allowed, with your application. They are brief, simply introducing you and your project and thanking the grantor for consideration. They also provide your contact information and Web site address. The question may arise as to who should sign the letter—the development director, grant writer, board president, or executive director? There's no best answer. However, if one of these people has a personal connection to the grantor, then the cover letter definitely should come from that person.

Summary

After a thorough review of your application, any one paragraph in your proposal could be the deciding factor that causes the reader to say, "This organization is not a fit with us" or "Let's support this organization." The thing is, we don't know which paragraph that is for every reader. Guerilla grant writers make every response to every question count. Our proposals are free of fluff. Our words aim straight for the jugular. Not one sentence or paragraph is wasted on irrelevant information that doesn't strengthen our appeal for support. And just as important, if guerilla grant writers come across questions they're unsure how to answer, we call the funder's grant officer to get clarification.

3

Preparation

As part of your research, take the time to read the guidelines to ensure your Agency's mission is a match with the prospective Company's focus areas of giving. Once that is clear that you are both in sync with each other, take it one step further to identify an advocate within the Company you are seeking funding from, as this could be a deciding factor in securing a grant.

—Mario P. Diaz, Community Affairs Manager, Wells Fargo Foundation, San Francisco Bay Region

You've heard about a grant opportunity, and you're eager to fill out the grant application or write a proposal. You've set aside a couple of hours, and you're ready to get started. Do you just dive in and begin answering the questions? Not yet.

This chapter identifies 11 crucial but uncomplicated steps the guerilla grant writer takes before writing the first words. Preparation is key, and it's a habit you want to acquire. The idea is to save time and energy; and completing these steps first will save you from frustration later. Remember, grant making is a competitive environment, and funders will weigh your grant application against others. These preliminary tactics give your proposals extra ammunition. How exactly? Well, think about this: a moving proposal, beautifully written and well-justified, will fall flat if sent to an inappropriate funder. A wonderful project proposal, greatly needed and widely supported in the community, will get a rejection if the applicant didn't follow the application guidelines.

In other words, no matter how perfect the grant proposal, a lack of preparation can lead to a rejection letter and heartache. But this won't happen to you. Once you embrace these steps, you'll wonder why everyone doesn't follow them. And when a grantor reads your proposal, it'll make sense and will be a viable candidate for funding.

Read the Grant Guidelines

Thoroughly reading the guidelines can save you from spending time on an application that has requirements you can't meet. For example, does the grantor require that you offer your services free to the public? Do they require sizable reporting on the ethnicity and income levels of your clients—something your organization doesn't track? Find out the types of grants the grantor awards and what purposes they support. Sometimes the guidelines will directly specify the types of organizations that can apply and the projects the grantor wants to fund. If you're not one of the eligible organizations, don't apply. Or if your project isn't an eligible project, don't apply to this funder with this disallowed project. Nothing you write will make them change their mind. The guerilla grant writer always makes sure that the grant request he submits fits the funder's guidelines for the grants provided. Additionally, the guidelines will give you an idea of the relationship the grantor wants with its grantees. Quickly weigh the pros and cons of applying, and make a decision as to whether or not to pursue further. Thoroughly reading the guidelines also will help you respond to all the steps that follow.

Determine if the Grant Is a Match with Your Mission

Finding a great grant opportunity isn't great if the funding organization's mission doesn't fit yours. You want to align your organization with mission-compatible organizations. Seek those that are a natural match. When their mission matches yours, they'll have an automatic affinity for you and your proposal, and there'll be greater potential for a long-lasting relationship. Look at it this way: while funders' grant making priorities may change to address societal or community needs, their mission won't change. Mission is a constant driving force. If your organization's mission ties to theirs, you'll have a bond that doesn't vary from year to year. If you've already established a relationship with that funder before their guidelines changed, that bond still may lead to opportunities to partner with that funder.

You may find most funders aren't a natural fit but rather are neutral to your mission. For instance, let's say that your organization is a library. A potential funder is a major oil company. The oil company's mission neither directly supports nor opposes your mission. No problem there. Many funders simply want to give back to the communities where their employees live. However, if your library is a green building model with LEED certification for meeting strict environmental practices in design and building, then you may look twice at the oil company and its practices. Does the company have a mission to decrease its impact on the environment? Are they supporting environmental organizations as a means of helping to improve the environment? At the least, seek funders that don't oppose your mission.

Estimate How Much Time It Will Take to Complete the Proposal

When is the application deadline? Do you have enough time to complete the research and the grant proposal by the deadline? Be realistic about the time the application and all its attachments may take to compile. Once you've written a few grant proposals, you'll have a good sampling from which to build future proposals, and those may take less time to prepare. But all applications are different, and each one still will take staff time. Another factor to consider is the time to prepare any reporting documents required if you receive the grant. Some grantors require extensive reporting, whereas others require none at all. Don't be afraid of required reports; just be aware of the time factor.

Look at the Giving History of the Grantor

You can find the giving history on most organizations' Web sites and in online annual reports. Sometimes you may have to call to ask for a hard copy of the annual report. Look for the types of projects the grantor funded in the past and what types of organizations they funded in the past. Are these similar to your organization or proposed project? How much did the grantor give? Do they award multiyear grants? Does the grantor tend to award the same organizations year after year, even in small amounts? While visiting the Web site, look for clues of other benefits the grantor provides to its grantees. Benefits range widely from providing a link from their Web site to yours, to providing volunteers to your organization, to major sponsorships of organizations with which the grantor has long relationships, to a lot in between.

If you come up empty trying to locate a Web site, you can look up a foundation's giving history through its most recent tax forms. Form 990-PF is available for free at foundationcenter.org. Type in the foundation's name and the state in which it is headquartered in the "Foundation Finder."

Guidestar.com also allows free viewing of Form 990 (see Appendix C). When examining the form, don't concern yourself with all the financials, but skim to the pages showing the grants awarded. Or just call the organization and ask. Although this step is one of the most important indicators of a match, you don't want to spend a lot of time on this step. Go to the most recent year first; then review the year prior and the year before that. A three-year history is enough. Look. Take notes. Come to a conclusion. The purpose is to determine which projects have a chance with this funder. Once you know that, you can stop.

Talk to Your Internal Project Leaders

Identify a suitable project from your organization's wish list or urgent needs list. Is the project a fit for this grantor? Ask for details about the project from your project leaders. Flesh out the idea with staff. Is the proposed project developed enough for you to write the proposal? Ask your project leaders to provide the projected costs, and ask how the project will be measured. Your organization's manager or executive director will need to confirm that the project will be implemented if you receive the funds. It is essential that your organization meet the parameters of the grant. Ensure that everyone involved knows the grant conditions before you apply. These conditions will be spelled out in the guidelines. They can be minimal, such as "Notify us when the funds are spent," or more involved, such as "Provide outcome measurements of the program or allow inspection of the capital project."

Call the Grantor's Representative

Introduce yourself, and run your proposed project by this person. Listen! Really listen! The representative may not come right out and say, "Don't send us that," but you can sense if he or she is not warm to the idea.

Forget shyness, and ask directly, "Is this project something you might support?" Ask any questions that are not covered in the guidelines (or Web site) or that you need clarification on. What is the average grant size? Does the grantor allow for staff and operating costs? (if this is what you need). Does your organization's location matter? (if it's unclear in the guidelines). Sometimes grants are location-specific. You don't want to take a lot of the representative's time, but you want to determine if your project is a fit for the grantor.

Decide to Go for It or Not

After your research and talking to the representative, you will need to decide if it's worth it to continue. You'll have an idea whether or not you can prepare a competitive proposal, the amount of funds you'll likely be awarded, and the benefits of establishing a relationship with this funder. Be willing to start small, particularly if your research indicates that the funder tends to fund the same organizations for several years—and more so if the grant awards increased over time.

If You're Proceeding, Gather the Information You Need for the Application

Your first grants will take the longest as you create or gather the background documents that are frequently requested. It's best to have these items handy in both soft copy for application proposals submitted online and hard copy for applications submitted through the mail. Items such as the organization's budget, accomplishments, organization history, and most recent financial reports may be requested. You also need the project budget. What is the total cost of the project? How much are you requesting, and where will the balance, if any, come from? You want to know (and clarify in the proposal) what these dollars will do. Determine whether you have all the background details about the project that you need to write a persuasive proposal. Are there any pieces missing? If so, ask for them now.

Review the Application Again Before You Start to Write

This will allow you to determine if another person in your organization has any needed data. If so, give that person a deadline to get the information to you. You also want the appropriate manager to review the project budget. Be sure that you are requesting the total amount needed to implement the project successfully. Asking for too much is a red flag. Remember, funders read thousands of applications throughout the year. They are experts at evaluating proposals and budgets. Asking for too little also sends up a red flag and may lead to disaster if your organization is awarded the small sum and bound to carry out the project underfunded. This application review also gives you a chance to ask the funder any follow-up questions. Calling the representative on the deadline day to ask your questions doesn't make a good impression.

Clear Your Schedule

Allow yourself time to write, review, revise, and assemble. It is always recommended to have another person read your proposal before you send it. Two readers are better than one. Allow time for this essential review by a fresh pair of eyes. Keep in mind that as much as you'd like to clear everything off your schedule, you'll likely still be performing some of your regular duties, so allow time for interruptions.

Find Out if Your Board Has Contacts with the Funder

It's not uncommon for a board member to have a day job at grantor corporations and businesses or to also serve on the boards of foundations. While you're busy writing the proposal, the board member can prepare to introduce and champion the proposed project to the grantor's representative. Many employers give preference to grant applications that are employee-recommended, whereas corporations and foundations may give preference to grant applications that are recommended by board members.

Now That Your Research Is Complete, You Are Equipped to Write the Proposal

Congratulations! The steps you engaged in give you confidence to proceed; the process involved others in your organization, ensuring that they support the project; it required you to think about your mission and give your mission a high priority; it required you to contact the funder and tell the representative about your organization; and it certainly will result in a stronger proposal than if you hadn't prepared yourself. See how much you've accomplished already?

Summary

These may seem like a lot of steps, but they are necessary to adding efficiency to the grant writing process. A little research up front goes a long way. Keep that initial eagerness throughout these steps and especially during the writing. A few hurdles don't thwart guerilla grant writers; we are prepared for them. Guerilla grant writers size up grant opportunities and pounce on those that ultimately complement our organizations.

4

Writing

Indiscriminate usage of buzz words such as mission statement, hands-on, underprivileged, diversity, etc. weaken a grant request and gives the impression that the grant request has been taken from some form book. ... Use the words that you would use to explain to your friend, your purpose, your needs, and your intended usage of the funds.

—Robert T. Borawski, President,
The Robert Brownlee Foundation

This chapter reveals the simplicity of writing a proposal. I'm even going to say up front that this book isn't about writing the perfect grant proposal. Perfection is not the goal. The goal is to get the most compelling, clear, and fitting proposal that you can write and get it out the door. This chapter will show you how.

Anyone can master these concepts by thinking in threes: beginning, middle, and end; short, medium, and long; and inform, persuade, and assure. In guerilla terms, that's kick butt, take names, and take charge.

Writing a proposal can be made easy by thinking of composing it in simple steps.

After five years of experience as a grant writer, I attended a philanthropy awards luncheon where I heard one of the speakers, Robert T. Borawski, say that his organization, The Robert Brownlee Foundation, only required the following in grant proposal letters: "Just tell me in two pages who you are, how much you want, and what you'll do with it."

The foundation's philosophy was for the grantee organization to get to the heart of its request (and to make the grant writing process as uncomplicated as possible on organizations). But I realized that the statement summed up what all grant proposals need. Breaking that statement into three manageable steps, those questions are:

1. Who are you?
2. How much do you want?
3. What will you do with the funds?

Even if you were to write a one-page grant letter, this is the minimum you should include, with a couple of obvious elements, such as thanking the grantor for its consideration and your contact information, of course. Now isn't that painless? Can you go to your desk today and write a one- or two-page grant request letter using this format? Yes. Even the busiest of us can manage that. This book includes examples of successful short proposals for you to use as a guide. But often funders have their own questions they want you to answer in your grant request letter. Or you sense that you really need to give background or supporting information to a funder who may never have heard of your organization before or your new project. Below I present three more ways of looking at the three easy steps to writing a proposal. All these steps are meant for you to use as a starting point and adapt as necessary. While grant writing is important work, it is straightforward, not complicated. Don't make it complicated.

Keep in mind, regardless of the length of your proposal, that you should not stray far from mission. Even in the shortest proposal, highlight how your organization and project directly fit their mission and goals. Also emphasize how your organization is different from others in the same field.

One way to think of a grant opportunity is as a short, medium, or long application. Determine what is required to apply for the grant, and size up the length necessary.

Short, Medium, or Long

A short proposal (one to two pages) could be used for a letter of inquiry, a sponsorship proposal, an in-kind request, and a clear-cut request from a funder who knows your organization and may have funded that project or program in the past. A medium proposal (about three to five pages) might be right for a grantor who knows your organization when the request is entirely new or the program is one the grantor has not funded previously. It's also used for grantors that need to know more about who you are and what you do. Many grantors' guidelines have specific questions for you to answer in your proposal, and the result will fall into a medium length. A long proposal (six or more pages) may be necessary when asking for support for a larger project (there's more to describe), for organizations that have never heard about you, and when the grantor's guidelines require several questions to be answered or the grantor wants more in-depth explanations.

Then, once you have the length, chop the proposal into the beginning, middle, and end. You can structure each of these three components as necessary for your length.

Beginning, Middle, and End

For example, the beginning of a short one-page letter will be a few sentences. The beginning of a long proposal can be a couple of paragraphs. The middle of a short proposal might be a few paragraphs, whereas the middle of a long proposal could take a few pages. However, regardless of the number of paragraphs or pages, the beginning, middle, and end must perform the same purpose. This means that the middle of a short grant proposal must deliver the same punch in those few paragraphs that took a few pages to deliver in the long proposal. So what are those purposes? Here's another set of three:

1. Hook the reader.
2. Inform, persuade, and assure.
3. Emphasize urgency, spur action, and close.

Address these three essentials in your proposal, whether it is short, medium, or long. As you can see, this concept does not conflict with the opening concept. It's just another way to look at breaking down a grant proposal into three steps.

The Beginning: Hook the Reader

Your opening will either grab the reader's interest or not. It should excite the reader and make him or her eager to keep reading. Ideally, it sets the reader's frame of mind to thinking, "*Yes.*" You want the reader to believe that he or she is about to hear of a very interesting project—one that is attuned with their funding philosophy. Hooking the reader doesn't mean exaggerating or promising the sky. It isn't phony. It's just the opposite. It's real and honest and readily understood. For example, look at how one organization opens its proposal for the three lengths:

Examples

Beginning—Short

"Our proposal presents you with the opportunity to invest in our Electronic Waste Recycling Program that promotes animal conservation and protection of the environment. An investment of $20,000 from the XYZ Foundation will enable ABC Youth Environment Center to implement our mission of engaging youth in learning about and preserving the natural environment."

Beginning—Medium

"Our proposal presents you with the opportunity to invest in our Electronic Waste Recycling Program that promotes animal conservation and protection of the environment. An investment of $20,000 from the XYZ Foundation will enable ABC Youth Environment Center to implement our mission of engaging youth in learning about, appreciating, and helping to preserve the natural environment. XYZ Foundation's support will allow us to develop and produce an educational DVD, launch a middle school campaign in Richmond County, and significantly expand our Electronic Waste Recycling Program."

Beginning—Long

"Our proposal presents you with the opportunity to invest in our Electronic Waste Recycling Program that promotes animal conservation and protection of the environment. An investment of $20,000 from the XYZ Foundation will enable ABC Youth Environment Center to implement our mission of engaging youth in learning about, appreciating, and helping to preserve the natural environment. XYZ Foundation's support will allow us to develop and produce an educational DVD, launch a middle school campaign in Richmond County, and significantly expand our Electronic Waste Recycling Program. Our project's goals are to increase the number of unwanted electronic waste products being turned in for recycling, thus reducing the amount of toxic waste going to landfills, and to inspire youths to become environmental stewards. ABC Youth Environment Center has formed a vast network of support from teachers, local environmental leaders, and parents."

The long proposal then might go on briefly to describe the urgent need to recycle, followed by a paragraph about how the problem affects human beings and nature.

The Middle: Inform, Persuade, and Assure

You want not only to tell the grantor who you are, how much you want, and what you're going to do with their funds, but you must, in those same sentences and paragraphs, persuade them that this is a much-needed project, that you have the capability to carry out the project, and that your organization is the best organization to be trusted with this project. Your words should give the grantor confidence in your project and make the grantor enthusiastic about partnering with you. In short, your words must do double duty. On whom will the project focus? What is the expected outcome for these individuals? Use the middle to educate the grantor. Reveal your organization's enthusiasm for and commitment to what you do.

For example, look at one element of the middle of the proposal (Who is the target audience?) for the three lengths:

Examples

Middle—Short

"ABC Performing Arts School has been serving teens in El Paso, Texas, since 1989, providing teens with opportunities for learning and personal growth. We serve about 300 teens annually, with 85 percent of participation by girls between the ages of 13 and 16."

Middle—Medium

"ABC Performing Arts School has been serving teens in El Paso, Texas, since 1989, providing teens with opportunities for learning and personal growth. We serve about 300 teens annually, with 85 percent of participation by girls between the ages of 13 and 16. Our students learn various dance techniques from ballet to tap under the guidance of experienced instructors and mentors. Through performance and roles in directing and choreography, students also gain self-confidence and leadership skills."

Middle—Long

"ABC Performing Arts School has been serving teens in El Paso, Texas, since 1989, providing teens with opportunities for learning and personal growth. We serve about 300 teens annually, with 85 percent of participation by girls between the ages of 13 and 16. Our students learn various dance techniques from ballet to tap under the guidance of experienced instructors and mentors. Through performance and roles in directing and choreography, students also gain self-confidence and leadership skills.

"We are the only nonprofit dance school in the area that provides scholarships for underprivileged students to obtain performing arts dance experience as well as our unique behind-the-scenes experience in choreography, directing, costumes, and makeup. These at-risk students lack access to performing arts opportunities in their community. Not only do our students stretch their creative wings in a safe and nurturing environment, but also they are learning an appreciation of the arts. A partnership with El Paso schools rewards students who receive a 3.0 or greater grade point average by allowing them to participate in the travel troupe for one week during their spring break. The troupe program is made possible by community support at no cost to the students."

In the preceding examples do you see that all you're doing is building on the details, giving more insight into your project? Whether short, medium, or long, the descriptions were strong. The longer lengths gave further important descriptions. The length of your proposal provides you with an idea of how much depth to aim for.

The End: Emphasize Urgency, Spur Action, and Close

The last thing you want the reader to do is to put your proposal down in a stack with others. You want the reader to do something with it. Your ending should be so compelling that if the grantor is a good match (and since you did your research, I'm assuming that you'll send proposals only to funders who are a good match), your proposal will go in a "follow up" stack or the "consider" pile, or the reader may slip a Post-It note on it, or the reader will send you an e-mail or give you a call. A compelling proposal should trigger some action on the part of the reader; add a compelling end, and you'll have a better chance that the action is a step closer in the direction of funding. Endings can be any length as long as they strike a chord. Here are three examples.

Examples

End—Sample 1

"As a sponsor, you will receive exposure on our Web site, e-newsletter, event press release, and promotion at school sites. Imagine the community exposure and goodwill for XYZ Business. However, to receive all the benefits of sponsorship, please contact me by January 1, 2011. If you have any questions, call me at [xxx-xxxx] or e-mail me at [xxx@xxx.com]. I will contact you next week to discuss this opportunity. Thank you for your consideration.

Sincerely, . . ."

End—Sample 2

"We invite the XYZ Foundation to join with us in making a critical lead gift for the opera house renovation. Together we can restore a significant structure for community use, economic development, and historic preservation. If you have any questions or seek additional information, please don't hesitate to call.

Sincerely, ..."

End—Sample 3

"Your investment will be much appreciated at this crucial time. As a thank you, XYZ Bank's name will appear in an upcoming issue of our newsletter and our annual report, and you'll receive recognition on our Donor Board (located at our lobby) for a year. We are grateful for your past support and hope we can partner again to brighten the spirits of hospitalized children and their families. If you have questions, please call me at [xxx-xxxx] or e-mail me at [x@xxx.com]. Thank you for your consideration.

Sincerely, ..."

Once you understand the simple foundation of a grant proposal, you can provide additional details, as necessary, by adding information from Chapter 2 and going into depth with those details. Here are four essentials to consider when building a proposal by adding components:

1. Take your facts and add heart.
2. Relate everything to the project you're seeking support for.

3. Keep mission in mind—yours and theirs.
4. Add your organization's personality, if it isn't already there.

A final note about grant proposal writing: Don't sound like you're begging. You're representing a respectable organization. And you're presenting a proposal to a grantor that allows that grantor to carry out its mission. The grantor needs *you*. So write with confidence. You have the knowledge, you have the understanding, and you have the fire within you for your cause.

Once you have the package completed and assembled, check items off your application checklist. Check everything twice. After all the effort from everyone involved, you don't want to ambush yourself and get rejected because of an incomplete application. Grantors receive far more proposals than they can fund, and they don't have time to call organizations for missing items. Have a colleague run through the checklist as well. After that, it's ready to send. Congratulations!

Summary

It's natural to look at grant guidelines and think, "This is too much work," or "I can't possibly finish this in time." Yes you can. Just break the grant proposal into manageable steps. Repeat after me: Grant proposals are easy and fun to write. In fact, you can write them in three effortless steps. Guerilla grant writers roll up their sleeves and dive in, tackling one step at a time. Guerilla grant writers build on the essential foundation of a proposal without fear. Our words educate potential funders and gain funder confidence.

5

Standing Out from the Crowd

It's about alignment—between your mission, your strategic plan, and the needs of your community. Don't write a grant for a project that is not mission critical. Make sure that your project reflects your own organizational priorities, and make sure that those priorities are in sync with your public service mission to serve your community.

—Marsha L. Semmel, Acting Director,
Institute of Museum and Library Services

Every grant writer wants to submit an irresistible grant proposal. Half of making this happen comes from doing the proper preliminary research. The other half comes from what you present on the page. Putting the concepts you've learned from previous chapters into practice is essential. But guerilla grant writers know that not all proposals are created equal. Some grant proposals that funders receive will clearly stand out from others. This chapter is about giving your proposal that extra advantage so that it's one of the ones that rises above the pack.

Below are some simple but advanced tips for writing a grant proposal. Look at this list before you write your first proposal. Once you've written your proposal, run through this chapter as a final review to add extra punch.

Easy to Read

Is your proposal clear and easy to follow? Does it flow naturally? Bullet points are great for providing white space on the page and organizing facts. Straightforward headers with bullet points make it effortless for the reader to find information. Don't be afraid to use the grantor's questions as headers or to organize your responses with your own headers. See Part 2 for sample grant proposals, and note how the narratives were organized. Follow the grantor's instructions on the word and page count of your submittal, and you're off to a good start. Many funders have online applications where the look isn't an issue. Instead, you'll have word or character count limits.

Warm Style

You want to write in the active voice, and you want a personal tone. Think of the active voice as actions you see taking place; think of the passive voice as vague thoughts. Communicate the impact this grant award will have not only in numbers but also in lives touched. Who will benefit? Make a powerful statement by personalizing the impact. Instead of "Fifty children will be reached with your support' (vague), make more of an impression with a clear picture of the lives touched: "With your support, 50 children will have one-on-one mentoring with role models, gaining self-confidence and learning lifelong leadership skills" (active voice). Personalizing your numbers sounds like you care about what your organization does.

Useful Content

Be sure to give grantors what they want. Always answer all the required questions. All the information you provide helps funders decide if they should award you a grant. It's tempting to include all the awards you've won and your long history of success. But you risk losing focus on the questions the funders want answered about your project. You want to provide useful and relevant details that relate to the grant project, answering any questions that may arise as the grantor's representative reads, erasing any doubts about your project or your organization, and reminding the representative why you are the perfect organization to partner with them. Also, clearly state what you want from the funders and convey how your proposal will help them fulfill their mission.

Know What to Emphasize

Some details are more important than others. Some facts need to be underscored, whereas other specifics should be left out of the proposal. An irresistible proposal is free of extraneous information. In other words, don't drone on and on about a topic—get to the point. You can emphasize without putting the reader to sleep. Often fewer, well-selected words do more to call attention to your points than heavy-handedness. The grantor's guidelines will help to determine what can be highlighted. On what does the grantor put emphasis?

Track Record

If the grantor has funded your organization previously, remind him of his past grant's impact. Assure the grantor, through your relevant past accomplishments, that your organization can implement the grant funds as you've stated and that the proposed budget is realistic for this project. Any similar projects or programs you've completed will help to show your ability to be successful again. Demonstrate your gravitas. If your organization is new, perhaps it's the executive director, staff, board members, or volunteers who have the successful track record for this type of project. Or possibly you're collaborating with another community organization that has experience in this type of project.

Reputation in the Community

What does the community think of your organization? Is your organization highly regarded? Write it down. Don't assume that the readers of your proposal know about the impact you make in the community. Persuade them that you are a great organization with which to partner by conveying a solid reputation for performing your mission well. Your mission is why you exist, and believing in your mission is half the reason a grantor awards you a grant; the other half is that the grantor believes in you. It's not enough simply to say that you're benefitting your community. Tell the grantor what difference you have made and to whom.

Your Organization's Personality

If you manage a women's shelter and provide a safe, caring environment for your clients, your descriptions should reflect that safe, caring environment. If you operate an after-school clubhouse for teens, your descriptions should reflect the creative, active, encouraging atmosphere of the clubhouse. Through your word choice, you reveal not only your mission but also *how* your organization performs that mission. Even the statistics and accomplishments you choose to share can say something about your personality. Maybe you have a distinctive way you recognize your donors (a way only your organization can)? This shows personality. Another way to reveal personality is to describe what makes your organization unique. Is your organization one of a kind in your area? How?

A Sense of Urgency

Clearly state a need in your community. Spell out why this need is urgent. Then explain how your proposal is the solution to the problem. Clarify why your organization is best suited to solve this problem. Why should the grantor fund your proposal over another proposal addressing the same problem? Urgency is a serious matter in a grant application. Why would your organization take time and energy to provide this program or project if it weren't a high priority? Why are you addressing this problem right now? You want the reader to get the urgency of the issue. You want the reader to feel compelled to help immediately, not the next grant round. Describe the urgency in personal terms, not just numbers. It's not just the number of cans of food you need but also the number of people who will go to bed hungry. It's not only the number of veterinary surgical units you need but also the number of cats that are euthanized each year due to overpopulation. (Of course, you provide the hard numbers in the budget section, too.)

Inspire Confidence

Grantors receive many more applications than they can support. It's a competitive process, and the grantors want to make the best choices so that their funds make a difference. You've told them how their support of your organization will make that difference; now you need to convince them that your organization will spend the grant in the way you claimed and that the outcome will change lives. Instill confidence in your organization's ability to meet the expectations of the grant project successfully. First, you need to believe it before you can write it. Write with authority, not meekly. State your organization's strengths. Highlight the goals of the project and measurable expected outcomes.

Present an Opportunity

Ideally, your grant proposal should come across as an opportunity not to be missed. In fact, it may seem as though you're doing grantors a favor bringing this opportunity to them. This is what happens when your proposal allows grantors to fulfill their mission at the same time it serves a need in society. Tell grantors why you chose them. Is it because of their business philosophy, their employees' use of your facility, or their granting priorities? Does their mission coincide with yours? Help grantors to see their natural connection to your organization and your project.

Pull Out All the Stops

Sometimes you just know you need that little something extra. You want one more punch that will set your proposal apart from all others. One way to add that zing is by referring to experts. Include mention of studies performed by your industry, your peer institutions, the government, etc. and how those studies reinforce the need for your project. If your organization has hired consultants, by all means refer to their findings and recommendations that relate to your grant request. Has your organization performed internal studies? Do you have a strategic plan, master plan, or long-range plan? Briefly point out the section that correlates with your proposal. All these prove that your organization isn't submitting a low-priority project and that you have thought about the problem your project is addressing and other experts have too.

Compelling Quotes

Do you have letters of support or thank-you letters from your clients? While many grant applications don't allow you to add them as attachments, you certainly can pull some of the text out and add it to your proposal. Pull the heartstrings with testimonials that tell the story of the lives you've touched. An appropriate quote that paints a persuasive picture can be woven into almost any response. (If you don't have letters from those you serve—schools, shelters, other agencies, and the like—you definitely should ask them to write one. What organization wouldn't want to help you get a grant that allows you to provide it with services?)

These basic tips will add power to your proposal, and all they add up to is putting into words (and numbers) what you've been doing, what you propose to do next, and the value of your organization to the community. Yet notice the energy these ideas bring to your proposals? Using these guerilla tips as part of your grant writing regimen will give you the secret ingredients that add vitality to your proposals and make your grant applications irresistible.

Summary

Guerilla grant writers set our grant proposal apart from the pack. We are proud of our organizations' strengths and achievements, and we articulate them on the page. We state our needs and our statistics in personal and meaningful ways. We make a connection from our organizations to the grantor's, making our proposal seem like an obvious fit and a great opportunity. We create urgency in every grant proposal we write.

6

Researching Grantors

I've had the opportunity to write grants and review them. Both tasks have the same goal: to further the mission of the nonprofit and funder with each grant. (June 29, 2010)

—Meredith Dykstra Hilt, Executive Director, Tellabs Foundation

This chapter describes guerilla secrets of where to look, what to look for, and how to approach potential grantors or partners. (While I'll admit that a few of these tips are not considered "best practices," they are how I've located some funders.) Would-be grantors are all around you; it's your job to find them and reel them in. The right grantor is one with whom you'd like a continuing relationship—not just one-time cash. It takes time to acquire grantors and sponsors; it only makes sense to use that effort on organizations that have long-term potential. The ideal grantors are keen about your mission (that gives you common ground from the get-go). As you're searching, keep in mind the characteristics that are important to your organization (e.g., mission-friendly, local, green philosophy, and industry-friendly, to name a few). Every organization has potential funders that are dedicated to their industry, such as the arts, the environment, animals, education, health, and so on. Thus, a history of giving to your industry could be an ideal characteristic for which you search. This is not to say that you disregard those that have never given to your industry or that don't meet all your characteristics. Rather, these are starting points for you to help narrow your search. As opportunities arise, your organization will need to evaluate whether or not to pursue them. This process could take five minutes or five weeks.

Perform Research

Look at New Businesses Coming to Town

New businesses want to build a presence in the community. They want as much positive exposure as they can get, and they only have so many advertising dollars. They want to reach their primary audience, and they may not have decided which avenues to use up front to reach that audience. Whether the business is privately owned or attached to a corporation, it may be willing to discuss a potential sponsorship of an organization that presents the right kind of opportunity. If you are a high school newspaper, located right in the neighborhood of the new business, your opportunity may be to include an ad for the business in the paper (if the business is beneficial to high school students, of course). The size of the ad and number of issues are things to consider. This goes with the discussion of what value you bring to a sponsor. If the business is a teen clothing store, a hamburger restaurant, or a skateboard shop, guess what? You bring a whole lot of value to the table. Their key audience is teens. Write a letter offering a sponsorship opportunity for $500 or whatever amount it is that you need to fund an issue of the paper. Then follow up with a call. (See the sample sponsorship letter in Chapter 14.)

Let's say that an established airline is coming to your town for the first time. What opportunity could your organization offer it? And what would you ask of the company? While you may prefer a cash sponsorship, maybe you have a fundraising auction approaching. Instead of cash, you could request a roundtrip air voucher for two. This offer is more inviting because it doesn't cost the airline anything, and the company knows that you'll eagerly promote the prize, thereby promoting the airline. Be aware of new businesses coming to town. However, businesses celebrate their grand opening only once, so you have to move quickly, before someone else does.

Look in Your Organization's Historical Files for Relationships that Have Slipped Away

For any number of reasons, your organization can lose touch with former supporters; those might be former board members and volunteers or former funders and sponsors. Hopefully, you have some sort of files that can be combed for contact information. If you come across a former grant application, note the date, name of the organization, name of the person to whom the application was sent, and the project and amount. See if you can find evidence of the grant being awarded. (Asking staff might be faster than searching through old files.) An Internet search will inform you whether the organization still exists and still awards grants. If so, reestablish the relationship. Being a former grantee is a bit of a bonus. A call to the current grant officer (if the one named in the application is not available) saying something along the lines of "XYZ Foundation provided us with a grant six years ago, and we just wanted to thank you again and let you know the classroom tables and chairs are still being used today."

Pause for the officer to put his teeth back in. That's right; he'll be shocked because no one calls years later to say this. "We'd like to invite you for a visit to see one of our fabulous classes in action and see if there's a potential to partner for one of our current projects." Even if he can't attend, the connection is reestablished. The response might be to reapply through the official grant channels. That's fine, and you can run your potential project by him to find out if he agrees that it is a competitive project. Add this person's contact information to your database and mailing list, and of course, add the grant application to your grant matrix.

For individual supporters, volunteers, and board members, verify whether their names are already on your mailing list or in your database. If not, give them a call. After you introduce yourself, you might say, "I noticed that your name wasn't on our mailing list, and I apologize that you haven't been kept in the loop. If we can verify that I have the correct mailing address, I'll add you immediately so that you can be sure to get future notices of happenings at our facility." If you don't have an e-mail address for them, ask for it. Then add the person unless he or she requests not to be added. (There also could have been an upsetting event that soured them.) Former board members, volunteers, and donors have an affinity for your organization. If their names have slipped through the cracks, this means that no one has asked them for a current donation. They haven't received your annual appeal mailings, they haven't received your newsletter that may have included a donation appeal, and they haven't received your love (in the form of invitations to special events, Christmas cards, and preregistration to programs if you offer such things to your donors). These forgotten supporters most likely have been sending their donations to other organizations!

Treat sponsors as somewhere in between grantors and individual donors. If they are corporations, complete an Internet search to determine whether they still actively award sponsorships or grants. If so, add them to your grant matrix, and make a note of your findings. The call to them will be different if they funded an event that is long over. If you're interested in having them sponsor another event, it will help to have statistics on the earlier event or like events before you call. The conversation might begin something like this: "We appreciated your support of our fun run five years ago. Since then, it has grown to 800 entrants and 2,000 attendees, and we have media sponsors and get coverage from local newspapers and TV. When you sponsored our event, we displayed your banner for a week and thanked you in the fun run program. Can we discuss the great benefits we offer our sponsors now?" Of course, if you ask this question, you must have great benefits to offer, or else ask a different question, such as, "Would you like to know more about our current sponsorship opportunities?" The former sponsor may ask you to mail the information. This is when you'll draft a sponsorship letter if your organization doesn't already have a sponsorship package developed. If the former sponsor was a small business, you'd want to verify that it is still in business. Your call to might be: "We appreciated your support of our holiday festival six years ago, and we wondered if you're interested in gaining the massive exposure of sponsoring another event this year." Use your own words to describe your sponsorship opportunities. Don't say *massive exposure* if you can't ensure it.

Examine Like Organizations

Peruse their ads (for upcoming events), Web sites, newsletters, and annual reports. Who are their sponsors and grantors? These are potential sponsors and grantors for your organization. Since they are enthusiastic about the like organization's mission, there's a good chance they'll have enthusiasm for yours, too. You're not looking to steal them but to introduce them to your organization and give them the opportunity to engage in your mission as well. You can be one of their grantees without taking them away from those they currently fund. While you're out in the community, visit local sites, and check out their donor walls. You may see the names of foundations or corporations you hadn't thought about approaching.

Assess Your Local Area. Who Are the Big Corporate and Foundation Players?

As mentioned in Chapter 1, corporations like to support the nonprofits in the areas where their employees live. They do this to improve the quality of life of their employees. They support organizations that help the communities thrive or that address social concerns, thus keeping their employees living in the area and working for them. Look at their ads. Sometimes you'll find a clue to what they might be interested in funding. For example, does an employment ad for a technology company stress the company's sustainability practices and commitment to the environment? Who should immediately research this company's grant giving procedures? And who should contact the company's community relations officer anyway, even if the company doesn't have an official grant program?

Community business publications and community newspapers may uncover promising leads to foundation or local business funders. See if there's an announcements section that may list grant awards or new business ventures. Is there an awards page announcing community recognition of an organization or individual for philanthropic giving? Did the organization or person make a gift to an organization like yours? Do these publications publish an annual list of the top foundations, corporations, or industry leaders? Is there a society section that shows photos of foundation CEOs attending nonprofit events? Are those nonprofits like yours? Does your Chamber of

Commerce have a publication or Web site with these announcements? Also scan trade newsletters and magazines for announcements of grants. Industries such as museums, zoos, and libraries have magazines for their trade. Past funding is the best indicator of the funding interests of the grantor. When you find a match to your organization, pursue it.

Attend Funders' Conferences or Workshops

Using the conference brochure, do a little bit of preliminary research on the speakers from grant making entities. Seek those who are a potential grantor. Look at their Web sites and see if you have any questions. When you attend the workshop, sit in front. Before the session starts, if the grantor isn't busy with the moderator, go up and introduce yourself, give her your card, and ask for her card. Yes, this is bold. Don't worry; before anyone else gets the nerve to do the same, the class will have started. If you have a good question (think up one beforehand), ask it during the class. Another reason to sit up front is so that the grantor sees you smiling at her, has eye contact with you, and remembers you. Also, if you have a further question, you're in the best spot to reach the speaker after the class to ask it. A funder you meet in person is likely to answer your questions more candidly than the standard answer such a funder would give you over the phone. Also, sometimes handouts with the grantor's most recent guidelines are available. Be on the lookout for these while at the conference.

If you can't attend the workshop, you still can make headway. Study the brochure. Usually speakers from grant making entities don't speak unless their organization is open to receiving grant proposals. The brochure may state valuable information, such as what types of grants the grantor makes and how much it gave last year. If there is a funder you really wanted to hear speak, find out if you can buy the tape or CD.

Put Feelers Out

Advertise Your Organization's Needs and Sponsorship Opportunities

You never know how prospective grantors may hear about your organization, but if something sparks their interest, rest assured that they'll go to the Internet to find out more about you. Their first visit may be to your organization's Web site. In addition to showing how your organization is fulfilling its mission, wouldn't it be nice for a funder to see sponsorship opportunities and some needs listed? Possibly you need a projector for your classroom, sofas and chairs for your women's shelter, a new animal exhibit at your zoo, or an updated sound and lighting system for your theater. Consider identifying your needs and sponsorship opportunities on your Web site, in your newsletter, and through social media. The funder, whether corporation, foundation, or individual, may call you for more information. Your organization's friends, such as members or volunteers, may see a particular need announced in your newsletter or on your Facebook wall that pulls on their heartstrings and spurs them to contribute to the cause.

Let the Public Know You Accept Donations

Thank Grantors, Donors, and Sponsors on your Web site, in your newsletter, in your annual report, as well as on an onsite donor wall so that others see they can have their names there too. Some donors don't connect your organization to one that needs donations. (This is especially true if you've never asked them to make a donation.) When they see you thanking others for their contributions, they make the connection, and ideally, they make a mental note of it. But you still have to ask specifically.

Where Do Your Board Members Work? Where Do Your Volunteers Work?

An often-overlooked source for finding untapped grantors is right under your nose. How many board members and volunteers does your organization have? Let's say that you have 50 total, and let's say that half of them are retired. This is still 25 potential companies where you have an insider employed. They can tell you about the company; that is, are they a good match? They can scope out the philanthropic climate for you. They can find out contact information. And they can rally for you to get support. Frequently, organizations let their employees select the organizations to support, and the company makes an outright grant (through the application process) or the company matches the employees' gift. You may want to add a question on your volunteer application and board application that asks about employment and other volunteer service. It's about making connections. Maybe someone has an "in" at the company to which you've been trying to introduce your organization. But you won't know unless you ask.

Do Recipients of Your Services Have Contacts?

It may seem uncomfortable to ask your clients, students, or patients for contacts. But often the recipients of your services or their parents (if the recipients are children) are more than willing to help. Put the feeler out in a manner that works for your organization. For example, a summer camp organization may wish to find sponsors or grantors to support the many camp sessions it conducts throughout the summer. A flyer handed out and an announcement made to parents at the camp orientation might yield the following: eight parents who work for six major corporations in the local area, two parents who work at banks, one parent who works at the local paper, and one parent who works at the local department store chain. That's nine potential grant contacts and one media contact! Some grant applications actually ask if any of their employees are involved with your organization! For this reason alone, you needn't be shy in asking about employment or other contacts. If any clients or parents offer a contact, always ask those clients and parents if you can include their names on the grant application.

Ask a Current Funder If It Knows a Cosponsor that Would Be a Good Match

One year I received a call in response to a sponsorship letter I'd mailed to an esteemed sponsor of an annual event. The caller said this: "We're excited to sponsor the event again this year, but I'm afraid we can't provide you with the same amount as we have previously." I quickly thanked the caller for the sponsor's continuing support of our organization and told him that we appreciated the company's sponsorship at whatever level it chose. But the caller went on to say, "Well, if you think it's okay, we'd like to ask one of our partners to cosponsor the event with us." It was my turn to be shocked. I hadn't seen that coming, and I wished I had thought of it. The sponsor's partner was one we never could have reached on our own—it was big-time. I said yes. The lesson here is to think outside the box once in a while. Use a little creativity when faced with challenges. Again, this occurred after years of relationship building with this sponsor. You wouldn't ask a new funder this question. However, if the situation presents itself with a long-standing sponsor, and you know it has a connection with another potential sponsor, inquire about it. "Do you think XYZ Corporation might want to be involved in the event?" A question as simple as this opens the conversation without offending the sponsor or putting the sponsor on the spot. As the conversation continues, the sponsor may offer to call the partner. If not, and the sponsor is not sure of the partner's interest, your next question is, "Who do you think I should call to inquire?" In this way, you obtain the contact's name.

Be Newsworthy

The more you get the word out in the community about your organization, the more grantors will see your name and hear about your activities—which will lead them and individual donors to your Web site. Make sure that your Web site has an easy way for donors to make instant online donations. (Your organization may want to use a service such as Pay Pal or Network for Good for online giving.). It may take several impressions (the number of times a person sees or hears your organization's name) before a potential funder decides to make a gift. Your marketing staff knows all about using press releases and other techniques to get publicity. You want to make sure that your Web site is donation-ready. Your organization also should think about announcing grant awards in your local business publication (on the free announcements page) or nominating a grantor for an award and announcing the grantor's selection in a press release, and you may encourage your board members to mingle at community events and speak at local clubs to talk about your organization. These tactics also get your organization in the news.

Online Searches

There are many online resources to help you locate potential grantors. Directory searches involve using a database. Some are free to use but give limited data about funders, whereas others may charge a fee for a fuller description. Basic online sleuthing involves using an Internet search engine (such as Google) by typing in key words to find grantors. Below are a few to try.

Directory Searches

Since these are databases, you have multiple "fields" by which to search. For example, you can search by mission, by geographic location, by type of funding, or by many other fields. You can even search by name, not only the grantor about which you are seeking information but also the name of like organizations to see who funded them, for what project, and for how much. What have you learned about funding to like organizations? (It can mean the grantor is predisposed to your mission.) Past funding data may not be available in the free version, however. A point to note is that organizations such as the Foundation Center provide their onsite database as a free service to local nonprofit organizations in San Francisco, New York, Atlanta, Cleveland, and Washington, DC. Perhaps an organization in your area does the same. Check out the list of possible sites at http://foundationcenter.org/collections/, and then call those near you to confirm what they offer.

Foundation Center

The Foundation Center's Foundation Finder and 990 Finder are free to use (http://lnp.fdncenter.org/finder/). When using the Foundation Finder function, you'll need the name of the foundation and state. This search will yield the contact information, the foundation's total assets and Web site address, as well as links to the foundation's most recent 990(s). For the 990 Finder directly, from the homepage of http://FoundationCenter.org, type in the word *foundation* and the year, select your state, and hit "Find." A list of foundation names will pop up. Click on the name of a foundation, and browse the 990 to see who the foundation funded in that year and the amount. The 990 also shows the foundation's address and board members' names. The Foundation Center's other online directory services are available for a monthly or annual fee.

Guidestar.org

Guidestar also has a search function if you know the foundation's name. You can search the foundation's 990(s) and read its mission statement, grant making program areas, and other useful information that the foundation provides. If the foundation does not supply the data, some areas may be left blank. On the first page, under "Search Guidestar," there is a field that says, "Nonprofit search." Type in the foundation's name, and click "Start search." This is a free service.

Chronicle of Philanthropy's Guide to Grants Page

The *Chronicle of Philanthropy* offers a free grant search of grants awarded (http://grants.philanthropy.com/pcgi-bin/premium/gtg/texis/grants/premium/gtgsearch). Type in a specific key word to search by type of grant, such as *animals, arts, health,* etc. Or you can search by grant maker or recipient (like organizations). Then you can further browse the results to find more leads.

Council on Foundations

The Council on Foundations has a Community Foundation Locator in its "Who We Serve" section on the community resources page (www.cof.org/whoweserve/community/resources/index.cfm?navItemNumber=15626#locator). Scroll down to the map of the United States, and click on your state. This will lead you to a list of all the community foundations in your state with contact information and Web site links.

Government Grants

The Grants.gov site may help you find federal grant opportunities (http://www.grants.gov/applicants/find_grant_opportunities.jsp). Browse by category to get started. Find your focus area, and go from there.

Don't forget to check out state grant opportunities from your state's official Web site. Each state will be different, but as an example, here's California's state grant offerings site: www.ca.gov/Grants.html.

When you find a potential grantor, visit their Web site for their guidelines and past grant awards. Follow the steps in Chapter 3. If the grantor's annual report is not online, ask for a copy. You'll see more about the grantor's priorities and the prior year's awards.

A final note about online searching: when using grant writing resource Web sites, make sure that the links on which you click are to real Web sites and not ads. Some links are misleading in that they are really ads trying to sell you something. Also, some book resources that list only a few sentences about each grantor may be too general or too dated to be useful. I've found that in addition to the grantor's own Web site and calling the grantor and asking, past giving is a top indicator of future giving.

Summary

Guerilla grant writers are not shy when it comes to our organization's mission. We are asking on behalf of the mission, not for ourselves. Guerilla grant writers aggressively pursue the right grantors and do not guarantee anything our organization might not be able to deliver. Guerilla grant writers research like local organizations and, just as important, add potential funders we have researched to our grant application matrix. Pick the few strategies from this chapter that work for you. Don't try to use them all, or you won't have time left for writing the grant!

7

The Waiting Process

I have always considered the various terms associated with fundraising emblematic of a process that at its best is one that takes time, is based on building trust, and in finding common purpose. Consider these position titles: Advancement, Development. They are quite different than Grant writer or Fundraiser. They are purposely general and expansive.

—Elisa Callow, Planning Consultant, former Arts Program Director for the James Irvine Foundation, and Program Officer for the Ahmanson Foundation

Waiting is not guerilla-like. There are productive things to do after you've submitted a grant proposal to fortify your grant writing activities. Tackle an item from this list regularly, whenever you can squeeze a few minutes in, and you'll soon have a robust grant writing program.

Research Other Grant Opportunities

Research is a main ingredient of the grant writer's duties. This is how you keep fresh opportunities coming for new grant funding and discover opportunities for creating new friends of your mission. Through research, you find the organizations that best match yours and avoid wasting time writing grants to organizations because you heard they gave a big grant to a neighboring organization. Wasting time is also not guerilla-like. Your modus operandi, or MO, from here on out will be to apply strategically to the organizations from which you have a realistic chance of winning funds. You uncover such grantors by being a sleuth. Keep your ears and eyes open because clues to locating fitting grantors are everywhere. Start with those closest to you, and work your way out. In other words, start with your community, your industry, and your organization's friends. (See Chapter 6 on specifically how to research grantors.)

Send Out Another Proposal

Once you've sent one proposal, there's no reason not to revise it and send it out to another potential grantor. This new proposal might request funding for a different project or a different aspect of the same project. Your numbers were up to date in the proposal, so revising it for a new grantor won't take as much time. (But revise it you must.) You want each grantor to feel as though you wrote the proposal just for her. And actually, you will have. Individualize the proposal to reflect the mission, funding requirements, and core priorities of the new grantor, all the while being genuine and convincing.

Develop or Update Your Grant Matrix

If you haven't already generated a grant matrix, it's time. As you conduct your research, you want to track the results of your findings at the same time you create a course of action for future grant writing. Every time you uncover a potential grant source, add the name to your matrix. Once you confirm the validity of the potential match, either remove the name as not matching or add the missing details to the matrix (i.e., the upcoming deadline date, the funding range, and the project you think is the best fit). Personally, I keep a second page to this matrix where I list, in alphabetical order, the organizations that are not a match at this time, and I date it. I include any note that's pertinent, such as "Closed to new grantees for two years," or "Only seeking arts grants at this time," or "Grants open to San Francisco organizations only." This effort saves time when six months later a board member suggests you look into a grant with so and so. You can say, "I already did, and they're closed to new grantors" rather than spending time reresearching a dead end. Take a look at Chapter 12. The importance of this matrix is to let you see at a glance where you already applied, the results of those applications, and who's next. Keeping it current is an ongoing process. Make it part of your regimen. The reward is in knowing your grant writing to-do list for the year, not letting a potential funder's name get lost on a Post-It note, and keeping track of your grant writing efforts as they occur.

Gather Quotes

Quotes can add credibility to your case. The right quotes add power to your proposals and help to emphasize the points you're trying to make. Where do you get these quotes? From your evaluation form comments, from letters, and from e-mails or posts on your Facebook wall. You also may get quotes or testimonials by requesting them from your clients and users of your services or their families. Let the person know you want to include testimonials in a grant proposal and/or your collateral materials. If the person requests anonymity, by all means accommodate her wishes. Someone needs to cull your pile of evaluation forms and thank-you letters, even those written by children, for the gems that tell the story of your program or organization. Type those nuggets into a quotes file that you can borrow from as necessary. Have your staff keep on the alert for quotes they come across. You may decide to mail some of those thank-you letters to the grantor who helped fund the program. Use your judgment as to what is appropriate to send and whether names should be blacked out.

Ask for Letters of Support

Generally speaking, you won't get letters of recommendation or support from the community unless you ask for them. Why would you want them? I've seen few grant applications that ask for letters specifically, but you may be able to attach one. If attachments are not allowed, remember that you still may quote from your letters in your project detail or other appropriate section. Words of praise from others carry a lot of weight. They are often inspiring and timeless. Here's one sentence from a full-page letter:

If not for the generosity of this organization, the children living in our shelter would not have had the once-in-a-lifetime experience of going to camp for a memorable week away from their domestic worries.

Do you see how illuminating one sentence can be? This sentence, and one or two more, could be included in the "impact on your community" section of a proposal. Teachers are usually willing to help to have their class create thank-you notes for the donor of a program their class benefited from.

Many times this is not possible during the program, so it is done back in the classroom. Children write the cutest letters. Who can you ask for a letter?

Take Photos

Most grantors welcome photos of the project they funded, especially those that show people enjoying the results of the project. Photos help prospective grantors to envision potential projects and let them see your facility or target audiences. But not all grantors want to see photos with the proposal. Take the photos anyway. You can always add a line in your project description that says, "Photos available on request." You also want photos in general of your facility serving your audiences and doing what you do every day. In a crunch, these may be the only photos you have. You know how it is. You wait until the last minute to take photos for a grant proposal, and the day you need to take them is the day it rains, or the camera battery dies, or a colleague fills up the memory card with photos. Take those all-purpose photos at an optimal time, and save them electronically; also print a few out (make a photo file for them).

Brainstorm Creative Ways to Recognize and Thank Donors

Now here's a fun activity in which to involve others. As you may know, brainstorming means coming up with a load of ideas, and all of them are written down regardless of whether they sound outrageous or not. Even a seemingly absurd suggestion can trigger the perfect means to thank your donors. Unsurprisingly, if you can tie the recognition or thank you to your mission, it makes that recognition (and your organization) all the more special and memorable. For example, I heard this story a couple of years ago: an organization in a foreign country had just had one of its currencies become obsolete. At its donor gala event, the organization gave everyone an envelope with a thank-you gift inside. Everyone was instructed not to open the envelope until told to do so. Finally, the board president gave his thank-you speech, and he asked everyone to open their envelopes. As envelopes were ripped open, the buzz stopped abruptly as the crowd eyed the useless currency inside. The president went on to say that if they had kept their money two years ago, today it would be worthless. But because they had invested it in the organization, the new clinic was now a reality. Wham! How's that for hitting home the point that donors can trust this organization with their money?

As another example, a zoo might have an animal paint a watercolor for its top donors. How would they do that, you ask? They simply dip the animal's hoofs in harmless watercolor paint and gently coax the animal across paper spread out on the floor. The process is as much fun for the zookeepers as it is for the animal. Donors love things made by children, be it artwork, letters, or something else tied to the project. As you can see, these techniques are tasteful, appropriate for the organizations, and economical. Organizations have used plaques, unique certificates, naming opportunities, and a host of inventive ways to say thank you or give public recognition to their donors. Check what other organizations in your industry or area have done to recognize their contributors. Can you adapt any of those ideas to suit your organization? (Here's where networking with other grant writers and fund raisers allows you to ask what methods they use at their organizations.)

Answer a Question from Chapter 2

Chapter 2 introduced many of the questions that can be asked in grant applications. Visit this chapter once in a while, and pull out a really good question to which you have an eloquent answer. The purpose in doing so is to weave those wonderful words into future proposals whether the question is asked or not. For example, an application you just reviewed but that isn't due until next month gives priority to "encourage the playing and enjoyment of symphonic and chamber music." The application doesn't ask how you reach underserved populations, yet your organization makes your music programs accessible to underprivileged children. You certainly would want to include your reply to the question in your program background or project detail section. Chapter 2 may nudge you to recall other vital things about your organization that you take for granted.

Examine the 40 Developmental Assets Mentioned in Chapter 2 and Appendix C

Write one thing your organization does that fits one of these assets. Use their language to illustrate what you do. Of course, it would be great if you did this frequently, until you had several moving and impressive sentences to express what your organization does for children between the ages of 3 and 18. Another useful aspect of doing this exercise is that you're practicing writing with emotion. Including emotion is another way to humanize an otherwise impersonal or analytical proposal.

Create a File of Requested Material

After you've researched a few funders' applications, you'll have an idea of the kinds of attachments they request. Label a hanging file, and toss in a few file folders. Have one folder for each of the attachments you've seen requested more than twice, for example, 501(c)(3) document, organization budget, board list, audits, and so on. Then, as you need these documents, you won't have to go hunting for them. This file will help you to stay efficient by having all your necessary attachments within easy reach. You also want them within an easy click of your computer mouse. Obtain these documents electronically before you need them. You'll also want a file for your quotes, letters, photos, and any other attachments, such as media clippings, that you may want to include. Again, while you're waiting is the ideal time to scan these items. Don't wait until the last minute.

Create a "Ways to Earn Funds" List for Your Organization

Appendix B provides a sampling of fundraising ideas achieved through a brainstorming session. Do your own brainstorming to see if you can concoct fundraising vehicles that your organization can implement. Of a list of 25 possible ideas, you may find only one that is feasible and suitable for your organization. But you needed the 25 to come up with the one. And one is better than too many anyway. You've heard this numerous times throughout this book, yet here it is again. Even the way you fundraise should tie to your mission. For example, an environmental organization might collect electronic waste as a fundraiser. A "friends of the library" group might sell books.

Summary

Don't wait for money to fall into your lap! Guerilla grant writers keep proactive. We use the time between writing grant proposals to prepare for the next proposal. We work at ways to strengthen our proposals and we have efficient workspaces. Guerilla grant writers seize opportunities to be creative, especially when thanking and recognizing donors.

8

Handling Acceptance

Stewardship is about relationships, not only donor relationships but every person in your constituency should be stewarded on some level, whether they make a gift or not, because one way or another, they can all come to support your institution's vision.

—Veronica Patlan Murphy, Stewardship Director, San José State University

Congratulations! Your hard work has paid off, and your organization is awarded a grant. Do you relax and forget about the grantor until you need money again? Not if you want your organization to have longevity, you don't. More important, you don't want the grantor to forget about your organization. Make contact with the grantor a few times a year. Avoid being a pest, but touch base with your donors from time to time so that they begin to feel invested in your success. Receiving a check is only the first milestone of the grant writing function. Here's what guerilla grant writers do to grow their organizations.

Thank Your Donor

Write your donor a letter, of course. Several months may have passed since you applied and last talked to your contact, so you also may call, under the guise of letting the donor know that you received the check, but also as a way of saying hello, we're still here, serving the community. Be appreciative. Even if you get the person's voicemail, leave a brief message. Don't ask the person to call you back. Just say, "It's [your name], from [your organization]. I just wanted you to know that we received your grant check for our [your program]. Everyone here is thrilled to have [the donor company's] support. Thank you so much." This is short, to the point, and appreciative—it's a reminder of who you are and what you do.

Send a Congratulations Letter to the Internal Recipient of the Grant

This form letter notifies program managers that the funds have been received and informs them of the terms of the grant. It reminds them to keep track of expenses and to gather program statistics and anything else you need to write a final report to the grantor. It also tells managers by when the funds need to be used and for what purpose specifically they can and can't use the funds. Copy your executive director and the board treasurer (or whomever is responsible for tracking expenses) so that they also know what is allowable to ensure compliance with the grant requirements (see example in Chapter 13).

Recognize Donors

Give your grantors public recognition in your newsletter, on your Web site, and in your annual report. These methods don't cost you anything, and they provide you with another opportunity to put your organization's name in front of grantors when you mail a copy of your publication. You also can recognize grantors at your grand opening or unveiling of the new project. Coordinate with them ahead of time, before the invitations are printed, to find a date that works with their schedule. Then mail them an invitation with an RSVP. Even if they can't attend, they'll see their name or logo on the invitation, "Brought to you by …," and will appreciate the recognition. If they do attend, make sure that their organization is mentioned and thanked. Depending on the size of the grant and the size of the grand opening event, you may have an event agenda, flyers, posters, or a banner printed. Be sure to include the grantor's name or logo on those, too. Many grantors already have banners with their company's name on them. They know that your budget is limited, and most will happily provide banners to you. It doesn't hurt to ask.

Inform Your Board of Donor Support

It's important for your board to know who their funders are. Board members may run into such funders at charitable or business events, and it's ideal if they can introduce themselves and thank grantors in person for the grant. The true board ambassador even will be able to say what the grant was (will be) used for and the impact the program has on the community. Talk about building an investment in your organization! To help your board members do their job, keep them informed monthly of grants you most recently applied for, grants that were approved or denied, and grants that you are shooting for next.

Add Donors to Your Donor Wall

A donor wall is that "permanent" fixture often found near the entrance to a nonprofit that lists the names of certain donors. Which donors are listed depends on the criteria you select, criteria that are appropriate for your organization. The donor wall doesn't have to cost a lot of money, nor should it be difficult to update. The important thing is that it look respectable, not cheesy. Ask yourself, "Would I be proud to have my name on it?" Locate your donor board in a public place where everyone will see it, and ideally, it should reflect your mission and inspire others to give. While the fixture is permanent, the names may change. Decide how long names will be recognized. When you write the thank-you letter, you can include a sentence that says, "In appreciation of your generosity, your name will be added to our donor board, located at the front entrance, for one year."

Add Donors to Your Mailing List

Create a mailing list if your organization doesn't already have one. In this way, you have addresses handy to mail newsletters, holiday cards, flyers of your events, etc. Choose wisely. Remember, you want to keep in touch, not bombard your supporters with junk mail. Also, personalize invitations to special events to your major donors with a handwritten envelope and brief note inside. "Hope to see you there!" or "Call me if you can attend" says you care. It's crucial to spell their names correctly, and if you don't know someone's title, give him or her the highest title within the realm of reality.

Invite Donors for a Site Visit

Let your grantors see their investment in action. It's one thing to talk a good game; it's another for grantors to see your progress with their own eyes. Plan for the best time to show them what you do when you'll be doing it at your best. Whichever program they are funding is what you want to highlight in the visit. You'll want your executive director and a board member to greet them and conduct the visit. Let them talk about what your organization does for the community, how the grantor has helped you accomplish that, and your goals for the future.

Take Photos

Sometimes grantors just can't find time to visit, or they prefer to stay out of the public eye. You still can provide photos to let your grantors see their dollars at work. You can include photos with grant final reports and with future grant proposals. You will need permission to use the photos if you're including them in your annual report or newsletters, and sometimes the grantor will want to feature the photos in their reports. Your organization already may include a photo clause within the fine print of your program registration form.

Create a File Folder

Each grantor needs a folder. Keep your award letters and copies of your congratulations letters, thank-you letters, final reports, and other correspondence inside. It's a track record of your communications and serves as a reminder to keep in touch. When someone asks you a question about the parameters of the grant, a quick look at the congratulations letter will tell you. When you wonder if it's been a year since you received the last grant so that it's time to reapply, a glance at their grant award letter will tell you. Keep important incoming mail from the grantor in one place. If you e-mail your reports, you may want a computer folder as well, but print out a hard copy for your file folder.

Keep Track of the Project for the Final Report

Send a midway reminder to the internal grant recipient. Your program coordinators are busy people. Give them a reminder to keep track of the expenses and statistics for your final report. A quick note is all that's needed. Ask if they have any questions, if there are any problems, and if the funds will be used up by the end of the grant period. Sometimes the manager will ask if the funds can be used differently because of an unforeseen circumstance. It's better to know about that early enough to still satisfy the requirements of the grant. Not only do you have to stay in touch with your grantors, but you also must stay in touch with your program or project managers!

Apply for Another Grant

Most grantors allow you to reapply a year after being awarded a grant. Sure, it's only been 9 months, but by the time the review period arrives and award determinations are made, it will be a year. Therefore, if you have results to show your grantors, or if the program they've funded is over, and you have the stats to show its success, go for it. This is where all your efforts during the year come back to help you. You've kept your grantors interested in your organization, you've invited them to visit, you've sent them your newsletter that acknowledged them as well as informed them of your influence in the community, and now you have measurements of your program's effectiveness. Remember: all the concepts for applying for a grant are the same for reapplying.

Send Donors a Final Report

The final report is yet another opportunity to thank your grantors and remind them of the important work you do. Sometimes there is a formal report with specific questions; other times grantors just ask you to tell them how the funds were spent. Grantors usually want to know the outcomes, such as how many people were served, how many programs were offered, and whether the project met expected outcomes. If you have thank-you letters from program recipients, for example, teachers, visitors, or clients, send a copy with your report. If there were any media advertising or articles about the project, include them. Give specifics about the program or project, and you already know that photos are welcome. (See the final report example in Chapter 13.)

Nominate Donors for Awards

Sometimes a grantor's support makes an enormous impact on your organization and the community you serve. It may not happen the first or second year, but perhaps over a multiyear relationship. If the final report you've prepared brings tears to your eyes and you're a hardened guerilla, it's time to step up their recognition. Look for an industry award, a local award, a mayor's award, or a community award, or perhaps your organization can begin giving an annual award. The Association of Fundraising Professionals (AFP) provides awards annually as part of a Philanthropy Day celebration. These competitive awards are given at the local and national levels to recognize outstanding philanthropists in many categories. Even if your nominee doesn't win, you still can recognize and thank your donor creatively. You can design an award that only your organization can give, something meaningful to the program the donor supports.

Ask for a Letter of Recommendation

Know that a grantor's support gives you credibility for other grant proposals. Often you'll see a request to provide a list of your funders in a grant proposal. Once in a while, you'll run across a request for a letter of recommendation from a community leader. A supportive grantor, one who knows that you provide significant services to the community, that you can measure and articulate your outcomes, and that you do what you say you're going to do with its funds, can provide a strong letter for you.

Cultivate Your Donors

Donor cultivation never ends. It's an ongoing practice—just like doing sit-ups and running laps—that leads to a robust, resilient fundraising program. Larger organizations have staff whose only job is donor relations. In a one-person office, that job is your responsibility. It's best to include your board members in this important task, so include donor relations in their job descriptions, too. Make it everyone's job. Think of your grantors as your partners. Include grantors in ways that are appropriate for your mission and your organization.

Summary

Guerilla grant writers celebrate grant awards with the best of them. We are appreciative of our supporters and proud of our organizations. We take responsibility to ensure grant compliance. Every grant proposal we write is one step toward creating visibility and longevity for our organizations. Every grant check we cash creates a partnership with the grantor. Cultivate your grantors. Thank them soon and often.

9

Turning Rejection into Future Victory

For every organization, finding new revenue and keeping expenses down is paramount. The key to securing grant dollars is both making the right match with your organization's mission but also making sure that you answer all of the questions posed. Instead of grant writing feeling like a burden, it can really be an effective way to showcase your organization and get much-needed revenue.

—Barb Larson, CEO, American Red Cross Silicon Valley

As a guerilla grant writer, you will receive rejections. Learn that "No" means "Not yet." Remember, you can eliminate some of those chances for rejection by doing the research up front. You're too busy to work on a proposal that has no shot at getting funded. We want to spend our precious time on grants that have a real chance. But sometimes even the best proposals are declined. Rather than sulking, think of the rejection as an opportunity to increase your arsenal for next time. This chapter outlines what you can do so that the effort is not wasted.

Call

Talk to the grant program officer, and ask what the weaknesses were in your proposal. Ask how you could have made it more competitive. At first, the officer may be reluctant to pinpoint weaknesses, but if you stress that his comments will help your organization present a stronger proposal for this project, he's apt to take the time to go over comments the reviewers made about the proposal. Just listen, take notes, and ask questions if you need clarification. Don't get defensive. Don't argue that the raters misread something you wrote. If it's not obvious from the officer's comments, ask if you can resubmit the proposal after it's rewritten. (Sometimes the officer will come out and say that the project was not popular or the project was not competitive. In cases such as these, do not ask about resubmitting.) It's also possible that the officer will say that the proposal was very well written and well received, but it just didn't score high enough. You'll learn so much about grant writing from taking this extra step. And guess what else? You're creating a relationship with that project officer.

Find Out Who Got Funded

Read the blurbs of those who did receive grants. Many funders will provide a list on their Web sites of the organizations that received grants and the dollar amount awarded, project descriptions, and the organizations' names. Was your project on the same playing field, or was your project unlike any of those awarded? Was your requested amount off base from those awarded? What other things do you notice about who got funded? Did you know that using the wrong terminology could affect your score? Calling your proposal an *operating grant* instead of a *program grant* or asking for *staff costs* instead of *program costs* could cost you if the term is one the grantor doesn't fund. What were the project titles of those awarded grants? Should you rename your project? Ask if the title or description made a difference.

Send a Thank-You-Anyway Letter

We grant writers are used to sending thank-you letters when we get a grant, but not many send a thank-you letter if we don't get a grant. Be one of the few. This is another way of getting your organization's name in front of the grant officer. You talked to her on the phone, and a few days later she receives a thank-you letter. The brief letter should thank her for taking the time to answer your questions and say something such as "Although we're disappointed that we didn't receive a grant for our dance troupe project, we appreciate your consideration of our proposal, and we hope to partner with XYZ Foundation in the future. There are many exciting things happening at [your organization's name]. Please let me know if you or anyone from your board would like to visit our site to learn more about us and the work we do to serve the community."

Keep in Touch

Stay in touch with the grant officer and grantor organization. Add them to your mailing list. Send them flyers of your upcoming programs. Show them your program in action. If the officer accepts your invitation for a visit or to attend an event, congratulations! Ensure that your executive director and/or board members meet him, and show him around your facility.

Reassess

After your telephone conversation and review of the grant awardees, take time to reassess whether this grantor is still a good fit for your organization. As you know, every so often funders change their giving priorities, and occasionally, their preferences aren't clear in their guidelines, or their guidelines are so general that it seems they're open to everything. Now that you've done your homework, decide if you should add this funder back to your grant matrix for the next cycle you qualify to apply to. Allow me to say that sometimes it takes a couple of applications to break in. Sometimes it takes three. I've been awarded grants on the third try. This is why I say that the answer "No" actually means "Not yet." It seems as though some funders purposely want you to show them you really want to partner with them. You almost did it the first time, but not quite. Not yet. Grant writing is not for the weak-hearted.

Reread

Set aside your proposal for a while. Reread it with fresh eyes. Can you see weak answers? Are the points the selection committee made valid? Do you see how you can strengthen the proposal? This is an important step whether or not you resubmit it to the same funder because you don't want to send a flawed or weak proposal out to a new funder either. And it's part of your learning, too.

Reapply

You've decided that the grantor is worth a second try. Check the Web site or ask the officer if the grantor offers a grant workshop or informational meeting. If so, attend. You don't necessarily have to reapply with the same project (it may have been funded already). So start the grant process fresh (see Chapter 3). If you are resubmitting the same project, make sure that you address all the points and weaknesses the grant officer mentioned to you. And be sure to call to run the project by her, noting how your organization may have changed the project elements. Get the officer's input on fit. Freshen the proposal up a bit, and revise anything outdated with updated figures. You don't want the reviewers to think, "I already read this exact proposal!"

Note: Some grantors have a waiting period before you can reapply for a grant regardless of whether it's the same project or a new project.

Know When to Move On

There comes a time when you need to realize that a grantor isn't going to fund your organization. There's a difference between determination and wasting your time. Once in a blue moon, a grantor actually will say, "Don't apply. Your organization is not a match with us." But more often, grantors won't say anything if you don't ask and your requests get declined. This is why it's important to talk to the grant representative. Depending on the grant opportunity and the amount, the most I have given it is three shots. If three proposals are denied, I definitely move on. However, for almost all grant writing, I do not attempt three times. With some, you will win a grant award after your first proposal, others after your second attempt.

A few might be awarded on your third try. After touching base with the grant representative, use your judgment on when to quit. Is the grantor warming up to your organization? Are their comments more encouraging? Is the representative offering any advice? Is this grantor awarding any organizations like yours? These are all questions to consider before moving this funder from your grant matrix to your "does not match" list. Such a decision may come sooner than after three attempts.

Be Kind to the Grantor's Representative

Here's another thing to keep in mind about the funding organization's representative: just because things don't work out with his employer awarding your organization a grant, do not feel that you've wasted your time cultivating him. That grant officer may move on to another organization with matching priorities. As mentioned previously, his organization's priorities could change in your favor. Or, as in many industries, he may have colleagues at other granting institutions that he meets at industry events. You don't want him saying that you treated him badly for declining your proposal, do you?

Stay Motivated

As you write more proposals, your writing will improve. You'll find briefer ways to say things, and you'll have more statistics and more stories. You'll come across deeper questions in applications that require you to delve into your purpose, analyze how you operate, and articulate your impact. Even when your proposal is declined for funding, your writing it was not a waste of time. You've introduced your organization and your programs to someone new. You're getting the word out there in the community about what you do. Keep learning. Keep writing. Keep submitting. Don't give up. It's for your mission, remember? Besides, guerilla grant writers don't whine. They get even (even tougher, that is).

Summary

Guerilla grant writers take positive actions to turn a rejection into an opportunity. We gather facts and our confidence, and decide whether or not to reapply. Guerilla grant writers realize that "No" means "Not yet." With careful follow-up, cultivation, and submission of a fresh proposal, however, "No" can turn into "Yes."

10

Achieving Long-Term Success

As you work to advance the mission of your organization, keep an eye on the future. While it's urgent to receive funding for the current fiscal year, it's imperative to also consider what those needs might be in 5, 10, 20 years. Assure institutional perpetuity.

—Sandra W. Gresham, CFRE, Director of Philanthropy, Island Conservation

By now you've written a grant proposal or you are prepared to. Most of us, however, want more than just one grant for our organization. We want sustained and increasing community support. This chapter will cover some habits to adopt right now that will position your organization for initial and continuing grant writing success. It offers tips to facilitate the perpetuity of your organization—in other words, serving the community forever.

Don't Promise What You Can't Deliver

It's tempting to offer the moon to potential funders. You want the grant, and you know the organization is a perfect match with yours. Sure, this one stipulation is something you don't think your organization can pull off, but one item isn't that bad, is it? You know the answer. It won't endear you to this perfect funder if you say one thing in your proposal and then don't do it—especially if it is apparent that you couldn't have done it. Grant writers must adhere to a code of ethics. (This code of ethics is available at afpnet.org under the "Ethics" tab.) Not only is misrepresenting your organization unethical (and possibly illegal), but your organization risks ruining its relationship with this funder and destroying any future funding chances with it.

However, sometimes an unforeseen issue comes up that prevents your organization from carrying out tasks or hampers your ability to meet a deadline. In such cases, immediately inform your manager about the grant issue and come to a solution with those involved in the project. Then call the funder's project officer assigned to you. Be straight up about the situation, along with your remedy. On another note, it is better to underestimate your outcome and report great news of doing more in your final report than to overestimate your outcomes and report not meeting your goals. Try your best to give an accurate estimate of your outcomes, but lean to the conservative side. If, after doing everything right, you still don't meet your goals, you still can write a positive final report. In addition to the meaningful numbers of those you served, the impact the grant support made, and your evaluation, include the challenges you encountered in implementing the project, how you faced those challenges, and the lessons you learned. This shows that you're better prepared for the next grant.

Don't Undervalue the Benefits You Bring to Donors and Grantors

When you're starting out writing grants and sponsorship letters, you're eager to get your first grant award, and it's natural to want to offer everything in benefits to those supporters: logo recognition on the front cover of your newsletter, acknowledgment at a sponsored event in front of all your guests, or prime banner placement for a month. Resist for a moment, and reflect on the value you bring to the table. Your organization has value in terms of name recognition, community respect, and your link to your constituencies. By supporting you, the grantor gets its name in front of your constituents, it gets positive exposure in the community, and it is connected to your project's success. These are all extremely valuable benefits to any funder. What other benefits can you tie to your organization? Even if you're a new entity, you have something to offer supporters. Once you've identified those benefits, then you can decide what to offer potential donors for their level of support.

Cultivate Year Round, Not Just When You Need Funds

Grantors realize that staffs at nonprofits are busy people. There's the impression that we're underpaid and overworked. And for many of us, this is a true actuality. However, there's no excuse for ignoring your donors after they award you a grant. Your organization must find the right way to *steward* your current donors, which is the term used to describe how you treat your donors as if they are special compared with everyone else, showing them how they've made a difference with their investment and ultimately leading them to making their next gift. (And for a healthy fund development program, your organization also will need to *cultivate* prospective donors.)

Cultivation and stewardship do not have to cost a lot of money. Buying gifts is not stewardship. Rather, stewardship is inviting the donors of an art enrichment grant to the facility to see the children building sculptures or painting with watercolors. Framing one of those pieces for the donor is stewardship. Thank-you letters from the children, even if written in crayon with little doodles, is stewardship. Obviously, what works for one organization won't necessarily work for another. Every organization can have a unique approach to stewardship. As with everything in grant writing, when it's tied to your mission, it underscores why the grantor got involved with you in the first place. Try to find time to devise a donor relations plan. In this plan, you'll identify how your organization will cultivate and steward donors, who will do it, when they will do it, and what the budget is for it. Start small. A stewardship event can be as unintimidating as a house party at a board member's home with the string quartet several donors funded playing their instruments for those donors and their spouses.

Don't Bring on More Grantors than You Can Cultivate

I know this sounds crazy—is there such a thing as too many funders? Possibly, yes. You've worked hard to establish the relationships—don't let them disappear due to a lack of stewardship. Determine what it is your organization needs to do to keep building the relationship. Is it bringing on more board members to help? This is where a little brainstorming ahead of time, with even the simplest plan, will help you to stay ahead. An organization that receives a grant and then doesn't contact the funder until a year later when it submits another grant proposal will not endear itself to the funder. Strive to keep your grantors engaged with your organization by keeping in touch. Engage donors with a site visit, a photograph of the project along with a personal note, or a letter from a board member saying,

"I saw the children's faces light up when they unveiled the dinosaur exhibit for the first time." Probably the most important thing you can do, besides saying thank you, is to show your funders the impact their grants made.

Deepen Partnerships

Back when you did your grant research, you might have noticed that your grantor gave several thousand dollars above its listed award cap to another organization. The Web site says a $10,000 limit, but the grantor's Form 990 or annual report says that it gave $50,000 to a nonprofit. This unusual grant gift happened, most likely, because of the relationship the organization had established with the grantor. These kinds of relationships happen over time. It's probable that the first grant to that nonprofit was indeed $10,000. After years of relationship building with not only meeting the parameters of the grant and having a great impact in the community but also some stewardship, the funder may have decided to award a much larger sum or a multiyear grant of $10,000 each year for five years. Never take your funders for granted. Appreciate them. Relationships are deepened when the funder is shown how it's making a difference and is thanked often, and your organization gives the funder the opportunity to do more. Additionally, when you've received a grant award, the grantor trusts your organization will carry out the grant as promised. Keeping that trust builds the relationship.

Be Willing to Start Small

There may be a funder that you believe is a great fit for your organization, but the funder awards only $1,000 grants. Should you apply? In some cases, yes. You have to weigh the potential with this funder against the time to complete the proposal. A $1,000 grant now can lead to $2,000 next year and $3,000 the following year and may lead to a $5,000-a-year multiyear grant in the near future. A funder may award a small grant to see how your organization handles it (and the stewardship). Or the grantor may bring other nonmonetary benefits to its grantees. Also, an application for a $1,000 grant may take only minutes to complete, making it well worth the effort. Even a small grant gives you something to write on forthcoming applications when they ask, "Who are your other funders?"

Revise Your Grants. Don't Use a One-Proposal-Fits-All Strategy

Tailor the Grant to the Grantor

Personalize your proposal so that it seems as if you wrote it just for your grantor. The items to adapt are how the proposal fits the grantor's mission and funding priorities and anything throughout the proposal that relates to this change. Look for other wording that doesn't fit this funder as written. Make those modifications as well. Maybe the accomplishments you highlighted or the focus area you emphasized needs to be adjusted to better suit this grantor. Could those 40 Developmental Assets be tweaked or added to show how your organization is a match with the potential grantor? Ask yourself, "Why did I select this funder?" Whatever it was that made you see this funder as a potential partner is where you can start tailoring. "We are pleased to present this opportunity to you because. ..." Your alterations needn't take a great deal of time.

Tailor the Grant to Your Project

A proposal for curriculum development is not interchangeable with a proposal for playground equipment. You will be able to make use of parts of one such proposal, but much of it will need to be written from scratch unless you have another curriculum-development proposal from which to borrow. And if you did have an earlier proposal, that one will need to be updated to account for the program's growth, the changes to content, and other variations since the original proposal was written.

Tailor the Grant!

Proposals are read by people, and real people can spot a proposal that is so generic that the organization is obviously trying to write one proposal to send to everybody. Remember I mentioned earlier that relationship building leads to future funding? You're off to a bad start if the first impression you make is with a proposal that lacks any personalization and reads like a boilerplate.

Accept Invitations to Attend Grantors' Events and Invitations to Be Featured in Their Newsletters or Annual Reports

Sometimes a grantor will invite your organization to visit its site, which could be at corporate headquarters or one of its satellite locations. Say yes! Numerous things can be accomplished by attending, depending on what the event is. It might be a forum where someone from your organization speaks to the grantor's employees about what you do or an employee community giving drive where all the nonprofits the grantor supports staff tables allowing employees to ask questions before they decide which they will select for their payroll deductions, or it may be an informal picnic day for employees with food and entertainment, and you're there to meet employees and have fun, too. These visits are part of your relationship-building strategy. Your organization needs a presence at these events whether that presence is your board members, the executive director, or other appropriate staff. Get the details of what's expected of you so that you're prepared, and once you're there, be an ambassador for your organization. And of course, have fun.

Similarly, consider it an honor to be featured in a grantor's newsletter or annual report. When the grantor calls to ask you for photos or information about your project, please be accommodating. (Remember to get photo releases.) Any visibility your organization can get is a plus, but it is all the more valuable when that visibility comes from your funder. Everyone will see it internally, which helps to deepen your relationship. It gets your name in front of the grantor—and you want your name in front of the grantor several times throughout the year. Be equally gracious with a request to be featured on the grantor's Web site. Often, though, grantee organizations will be added automatically to the grantor's Web site under the "Grants Awarded" section.

Remember, You're a Representative of Your Organization to Your Funders—Be Courteous, Professional, and the Purveyor of Your Mission

It goes without saying that you always want to maintain a positive spirit when you're working, but especially so when dealing with your grantors. Do not divulge any problems your organization is having with board members, landlords, or anything of the sort. Relationship building does not mean becoming chummy. Stay professional, and remember that the grantor's interest is in your mission, as yours should be, too. Likewise, if a grantor's representative should inadvertently inform you of confidential information about the grantor, do not gossip about it at the next board meeting.

Don't Expect Continued Funding

Your organization did everything right: You excelled with your project, you provided a timely final report, and you cultivated the grantor. However, the grantor informed you that you need to reapply, like everyone else, for another grant the following year. What went wrong? Probably nothing. Corporations and foundations have official grant making procedures. Many do not allow multiyear grant awards. However, another reason not to expect continued support is that grantors sometimes change their grant making priorities to reflect emerging trends in their community. Your project for an arts program, though it is highly successful, may take a back seat to housing homeless families due to a downturned economy and resulting job loss. A final incentive: when you don't expect continued funding, you don't take your grantors for granted. Remember, there are always other organizations out there trying to woo your grantors their way.

Keep Your Organization's Accomplishments Updated

As you know by now, there are a lot of little components that add up to compiling a knockout grant proposal. Anything you can do to streamline the information-gathering process will make your job easier. Get others to keep you informed monthly or at least quarterly on their ongoing program achievements, and make sure that you're kept in the loop about any other organizational accomplishments. Then develop an electronic file, and update it regularly. It'll be ready when you write your next proposal.

Keep Learning; Keep Yourself Invigorated

Optimistically, you've found that grant writing is enjoyable. Grant writing connects you to your organization's mission and causes you to truly appreciate all that your organization does for people and the community. One hopes that grant writing makes you feel empowered to earn support for your organization. But you also may feel something else, something along the lines of pride or victory, knowing that your words inspired others. Equally important, however, is to keep yourself inspired. Celebrate all your grant writing milestones: writing your first proposal, receiving your first check, and being awarded a second grant from the same organization. Continue learning about grant writing. Take a grant writing workshop, attend a grant writing conference, or join a professional fund development organization such as the Association of Fundraising Professionals (AFP). If you can't afford any of these, you still may attend local educational meetings held by the local AFP chapter. For a nominal meeting fee, you'll meet other fund raisers, network and make friends, and learn the best practices of the profession.

Perhaps challenge yourself with writing for a different purpose or tackling a government grant. Eventually, you'll gain confidence as a grant writer, and you'll learn to stretch your grant writing muscles to motivate yourself and to earn even more for your organization.

Expand Your Fundraising Program Beyond Grant Writing

The more you learn about the grant writing function, the more you'll be exposed to other forms of fundraising. For many, starting out with sponsorships may be the next step. You may wish to start out small and dabble your toes in a fundraising event, or your organization may be ready for a formal membership program. Organizations with individual donors may decide to create an annual appeal program to ensure that those donors are asked every year to repeat their commitment of support as well as to find new individual donors.

The point I wish to make here is that a healthy organization diversifies its streams of income. Your organization's funding may come from foundation, corporate, and government grants—and such diversity is fabulous. However, for enduring vitality, do not rely solely on grants for support. A multitude of options exist for your organization to consider. Appendix B identifies some alternatives that may be appropriate. Do not spread yourself or your resources too thin. Stay focused rather than trying too many other methods. The last thing you want to do is confuse your supporters regarding what your mission is. Fundraising, while essential to survival, should not become your mission.

Think About the Perpetuity of Your Organization

We all want our organizations to thrive, even after we've moved on. The best way to ensure that the grant writing function is not disrupted if your job should move you to other duties is to prepare for that eventual occurrence.

Train Others

In an earlier chapter I mentioned the importance of having others proofread your proposals. An added bonus to doing that is that you'll have two to three other people connected to your organization who've become familiar with the organization's grant proposals. These people will have an idea of what makes a great proposal, what was confusing to read, and what needs further clarification. These readers are the prime candidates to train to write proposals themselves. As you would expect, their proposals will need to be proofread by two to three different readers before sending them out.

Create a File System

Create a system for tracking the grants you've submitted, those that were funded or denied, and those you plan to target next. Chapter 12 provides a method of tracking grant proposals. More than this, though, you'll want a file for each grantor that contains all your correspondence with that grantor, a file for all your submitted proposals sorted by year or other identifier that helps you to retrieve the proposals with ease, and a set of files for all the attachments that you know must accompany most proposals (i.e., your organization's fiscal year budget, your 501(c)(3) document, your list of board members, and your financial audits, etc.). Although many of your proposals (and these documents) can be submitted electronically, keep a few hard copies on hand. The more grants you write, the more files you'll create to keep yourself organized. You'll also create procedures, which you should type out for your successor (or simply the person filling in while you go on a vacation).

Create Enthusiasm for Cultivation and Stewardship of Your Grantors

Train others in the significance of stewarding your grantors. Board members already may have incorporated donor cultivation into their job descriptions. But staff may not be aware of the importance of taking time with grantors. Certainly, before a site visit, prep staff included in the visit on who the grantors are, what you hope they'll see at the visit, and any questions you think they'll be asked. For example, "This is a reminder that XYZ Foundation representatives are visiting on Tuesday morning. They supported us last year with $10,000 for our Healthy Oceans Program, and I plan to request $15,000 this year. XYZ Foundation's priority is exposing children to ocean conservation. The foundation's representatives want to see a Healthy Oceans class in session, and afterwards they may ask about how you develop curriculum and how you evaluate the classes." Cultivation is everyone's responsibility, from the receptionist who answers the phone to the president of the organization and everyone in between. You never know when a grantor may pop in unexpectedly.

Never Compromise Your Mission

As your organization gains visibility, you may be approached by businesses to partner with them. Some offers may include a cash grant. The decision whether or not to accept of course will come from the leadership of your organization, but it never hurts to add your voice to the discussion. You know what's coming next. Question whether or not the offer compromises your mission. Short-term dollars could hinder your organization's ability to capture long-term support, especially if current or potential funders believe that the association counters your mission. (There's the trust factor, remember?) If the business entity is very mismatched, you even may lose board members and staff who are dedicated to your mission. Most offers won't be so drastic; this tip is basically a reminder to keep your head in place as you gain success.

Summary

Guerilla grant writers strive for the lasting success of our organizations as well as our grant writing careers. Guerilla grant writers counterattack attempts to stray from the organization's mission. We help our organizations gain the trust of grantors and help to cultivate and steward those grantors with every interaction we have with them. We help to communicate the organization's value and benefits to potential funders and stay informed about accomplishments to update those benefits as may be needed.

11

Other Uses for Your Proposals

Grant writing is a key skill for volunteers and development professionals alike, giving them the opportunity to educate foundations, corporations and individuals about the important mission-driven work being done by public benefit organizations. A good grant writer helps clarify the needs of the clients (animate or inanimate) and the ability of the organization, with the generosity of the donor, to make a meaningful difference in the future outcomes in a specific situation.

—Honey Meir-Levi, Executive Director,
Ronald McDonald House at Stanford

After all your hard work creating that masterpiece, it's understandable that you don't want to just file it away. You can't really frame it either, so what else can you do with those beautifully structured sentences? Plenty. Is there any further use for those descriptions that sing and the accomplishments that bring a tear to your eye? Absolutely. As a guerilla grant writer, you'll use the grant writing process to help grow the organization by implementing the strategies and tactics throughout this book. Not only will these tactics save you time and earn you funds, but they'll help your organization to gain visibility and prominence in your community as well.

New Grant Proposals

Use passages from your proposal for other grants. While most grant proposals are not interchangeable, parts of them are. Maybe you can make use of only the "organization background" section for one new proposal, whereas the "project summary" works for another. Certainly, the organizational budget won't change for the fiscal year. The more grants you write, the larger your potential to recycle your words. However, be sure to use the freshest data. Don't reuse outdated information or phrases. Even your favorite descriptions will sound old after a while, and you'll find it necessary to give them a kick or retire them. Keep a soft copy of your grants on your computer (as well as a backup flash drive or CD in your office) to easily access the files and cut and paste.

Annual Report

A long proposal will have lots of valuable information to include in your annual report. Depending on the design and length of your annual report, you can include your program descriptions and that year's numbers, accomplishments for the year, photos from that year's events, and so on. Annual reports are one of the avenues we fundraisers use to thank our donors and partners. Donor information is something you should keep handy not only for grant applications. (Some grant applications ask about your funders.)

Be sure to recognize individuals, foundations, corporations, and community businesses that have given you support during the annual report year. If your organization has never compiled an annual report, why not create one using some of the information and photos from your proposal? You can produce an online version so that there's no printing cost. Annual reports allow you to give a more compelling story to potential grantors than their applications permit.

Newsletter Articles

A section of your proposal can be fleshed out into an article, for example, an article about one of your programs or a series of articles covering several different programs. A single article can highlight a set of achievements, a signature event, or any other suitable elements you addressed in your proposal. An art museum, for instance, might have addressed its education programs in a proposal. One article angle might be to cover the entire education program and all its offerings (just as you wrote it in the proposal). Another angle is to select one facet, such as the teen mentoring program, to feature. View newsletter articles as an opportunity to get publicity for your programs and events to attract new patrons as well as a way to show current and future funders the value of those programs. (You also can send them a copy of the newsletter.)

Fact Sheets

Fact sheets are a great tool for providing lots of information in digestible bites. Some of those stats you collected or dates you highlighted might find a home here. Any interesting tidbit that helps readers to figure out what you do, how you do it, and whom you serve can go into a fact sheet. And your proposal should be full of interesting tidbits. If someone has never heard of your organization before, a fact sheet will get that reader up to speed on your history, accomplishments, etc., all in one page or two. If readers knew of your organization, they might be amazed by all the things they didn't realize you did. In other words, your most impressive information goes on the fact sheet.

Program Brochures

Your program brochures should catch the attention of the reader visually, whether that's with a captivating header, fabulous photographs, or great quotes. All these are items you can pull from your proposal. Portions of your proposal text describing the program might be fitting for the brochure, too. As will testimonials and fun facts. Think of the audience for your brochure and what you want them to do, and that will tell you what inspiration they need. For example, do you want them to make a donation, register for a class, or come to a special event? Each of these brochures will have specific text to get your desired results. Keep in mind their attention span when designing your brochures.

Annual Appeal Letters

Many organizations have annual appeals, during which they ask the community or just their client base for support. Such an appeal can be in the form of a letter or a brochure or even just a page in the organization's newsletter on the Web site. Whatever form the appeal takes, it needs to drive readers to send their checks or online payments. Your meticulously worded grant proposal will have just the right sentences to pull and use in your appeal. Don't forget those heart-warming photographs in your collection.

Sponsorship Materials

What benefits does it bring the sponsor to partner with your organization? You may have a separate sponsor kit or simply modify your media kit to include sponsor materials. These materials will provide additional motivation for a sponsor to want to partner with you. A long proposal sometimes will include such information as awards or recognition you've received, media you've attracted, or the demographics of those you serve. Copies of headlines your organization has received and newspaper or magazine articles written about you, or even ads you've paid for (ideally showing your sponsors' logos), help new sponsors to decide how much bang they'll get for their buck in sponsoring your organization. But other passages from your proposal, such as current or previous community partners, can be inserted in sponsor materials, too.

Web Site

Your Web site is the face of your organization. It may be the first and only interaction potential clients have with you. Naturally, you want your Web site to be informative, but also you want it to encourage clients to use your services and support your organization. There are numerous instances where your grant proposal wording, statistics, and photographs can be mined for their appropriateness on your Web site. You did the research; you may as well let the world see the impact your organization is having. As with anything else, be sure to keep those statistics up to date on your site.

Your Form 990 Description

Your organization files a Form 990 with the Internal Revenue Service every year, and you need to provide descriptions of what you do anyway. No sense in doing the work twice. Did you know that many of your potential funders will review your Form 990 online before making a funding decision? This alone is reason to make sure that you have real, dynamic, and sufficient information in your form. To see your current Form 990 descriptions, go to Guidestar.org online. Type in your organization's name in the "Non-profit Search" window. Don't scream or faint when you read what's there. It's often a simple matter of registering on Guidestar or other such free service to update the information at any time. Perhaps tax time is the annual reminder for someone to review that description for any necessary updates.

Updates to Funders

One way to cultivate your donors is by keeping in touch with them throughout the year, not just when you want money and not just with a final report. You may have written a beautiful description of the impact of one of your programs, which the funder supported, in your proposal. If you fashioned that into a newsletter article, then be sure to send the donor a copy of that issue. Write a quick note saying that you wanted the funder to see the difference its donation or grant made or helped to make. Or you may just cut and paste those statistics—which are personalized and impressive—and send that required report in the form of a letter.

Ads

Just as with your Web site, there are numerous opportunities for using your words, quotes, and more in an ad, be it print, radio, TV, online, or other medium. (Remember, you can only use photographs if you have permission from the adults in the photos or parents of any children in the photos.) However, through the course of writing your proposal, you may have crafted a catchy slogan for a program or an appealing caption for a photograph or a memorable title for an event. Don't let those snippets of creativity die in your file drawer. Bring them back to life in your various advertising channels.

Press Kits

Press kits are the packets that you give to members of the media and potential funders or partners. There are a wide variety of collateral materials you can include in a press kit. Many of the preceding items can go in, for example, the fact sheet, newsletters, and program brochures, along with the appropriate contact details. For media, the purpose of the kit is to generate interest in your organization. Show the opportunities and programs offered so that your organization seems newsworthy. The media may quote directly from the pieces you provide in their stories. The press kit is only half of it, however; you still want the members of the media to call your organization's media contact person for an interview. (The media contact person even may want to use some of your wording in her press releases.) You may customize the kit for whatever it is you're promoting or to whomever you're giving it. If the kit is for potential funders or partners, you might swap out some items, such as a personalized appeal letter or fact sheets about the project for which you are seeking funds.

These added uses aren't meant to overwhelm you with additional work. They are meant to make the effort you put into writing the proposal go a little further. These suggested uses are likely materials your organization generates already. Reusing your grant proposal elements may help you to be a bit more efficient. At the least, such an approach helps you see that developing one of the preceding items is within your reach.

Summary

Not every word in a grant proposal can be used again, but many sections of it are meant to be recycled. You went to a lot of time and effort in your research toward creating a piece that will resonate with funders and get your organization grant dollars. It only makes sense to give those words new life where others can see them. Guerilla grant writers repurpose our grant proposals to other uses within the organization.

Part 2

Samples of Successful Proposal Templates, Cover Letters, Budgets, and Reports

12

Grant Deadlines Matrix

As a grant writer, you'll soon find that there are many dates to keep track of: grant deadline dates, grant report due dates, and the dates applications were submitted—not to mention the status of each grant and what program the funds were requested for. To manage all this information, use a *grant matrix* (Table 12.1). This matrix has six columns in a Word document. (You can use a spreadsheet if you prefer.). The columns are self-explanatory, except maybe the funding cycle. During your research, you'll discover how often a funder has granting cycles. Some have ongoing monthly, quarterly, semiannual, or annual cycles. Some are one-time-only opportunities. Some require a query or letter of intent (LOI) first. (The LOI is akin to a summary of the proposed project. It allows the grantor to determine if it has an interest in your project before requiring you to submit a lengthy proposal.) Use the "Funding Cycle" column to identify how often the cycle occurs. Also use this column to identify any LOI queries.

Use the "Application Due" column to write in the deadline date. At a glance, you can see what dates are approaching and what grants you've already submitted. This matrix also can assist you to plan your grant

Table 12.1
Grant Application Deadlines Matrix for Year 2011

Name of Company	Funding Cycle	Application Due
ABC Corporation	Annual	Submitted 7/22/11
ABC Family Foundation	Quarterly	Submitted 2/28/11
ABC Investment Club	Annual	Submitted 3/26/11
Bank I	Semiannual	Submitted 2/1/11
Bank of ABC	Annual	Submitted 6/1/11
Community Service Club	Online application	Submitted 5/1/11
Corporation III	Quarterly	Submitted 9/19/10
Corporation IV	Monthly	Submitted 4/2/10
Federal government	One-time	Submitted 6/29/10
Foundation II	Semiannual	12/30/11
Foundation III	Quarterly	Submitted 9/23/10
Local business	Annual	11/1/10
Private fund	Preproposal pkg.	Submitted 3/17/11
State of California	Query LOI	Submitted 8/6/10
Utility company	Monthly	Submitted 1/5/11
XYZ Bank	Quarterly	Submitted 7/08/11
XYZ Corporation	Monthly	Submitted 1/24/11
XYZ Foundation	Query pkg.	10/1/11

Program	Funding Request	Response
Education operating support	$6,000	Pending
Education staff	$25,000	Awarded $10,000
Museum sponsorship event	$2,000	Denied
Unrestricted	$2,500	Awarded $2,500
Organizational effectiveness	$100,000	Pending
Exhibit opening event	$50,000	Awarded $20,000
General operating support	$25,000	Pending
Gala sponsorship	$10,000	Awarded $10,000
Exhibit conservation	$50,000	Awarded $40,000
Auditorium upgrades	$100,000+	
Mammal science exhibit	$100,000	Pending
Garden renovation	$3,000	
Environmental education	$40,000	Denied
Mammal science exhibit	$100,000	Pending
Environmental stewardship	$1,500,000	Awarded $200,000
Education program	$10,000	Pending
Education supplies	$6,000	Denied
Youth mentoring	$30,000	

writing efforts by allowing you to spread out your intended workload throughout the year. For example, you don't want to have five applications due in one month and none the following month. Just because a deadline date is June 30 doesn't mean you have to wait until June to write the grant. You can apply in March or April. This gives you some control over your grant planning.

Once you identify an organization as a potential funder, add its information. (If grant writing becomes your focus, your matrix may be more than one page.) Here's a tip: When I am busy and can't get to all the grants by a certain deadline, I apply for the grants that offer only one opportunity this year. If I wrote "quarterly" for one grant and "annual" for another and they both have the same immediate deadline, I'll apply for the annual grant, and change the deadline date for the quarterly to the next quarter. (Of course, you should evaluate the urgency of each grant need as well.)

Once you apply to a particular grantor, change the deadline date by adding the words, "Submitted 12/1/11." This date is also a trigger that any grant report is due close to a year from that date if the grant was awarded. If the grant was not awarded, the date is a clue as to when you can reapply. In the "Response" column, write "Pending" until you have a reply, and then write either "Awarded" and the amount or "Denied." At the end of the year, you'll have a record of your grant writing activities! Then you get to start the process all over again with a new matrix.

13

Congratulations Letter, Final Report, and Cover Letter

The congratulations letter is a useful tool to inform relevant internal staff of a grant award. I created the letter to remind everyone of the purpose of the grant and to help ensure that the funds are used as intended by the funder. It also includes any relevant notes about the funder or grant reporting requirements. (You may wish to add notes to your organization's treasurer as to which account will receive the funds, depending on the grant purpose.)

This informal letter also allows you to announce to the executive director, your board president, your manager, and other relevant staff that you have a new funder or continuing funder so that they can set their cultivation and acknowledgment strategies in motion. Another advantage of this letter is that it serves as a reminder to everyone that someone out there funded your organization! Someone thought your organization's work was important enough to award you a grant. How cool is that?

Finally, this letter helps to keep all the details of the grant on one page for easy access. A copy is filed in the funder's file folder. Soon you'll have many of these letters to file.

Sample Congratulations Letter

Congratulations!

A grant proposal submitted for your program has received funding!

Project: Education Program Supplies and Laptop

Date: 9/26/11
Amount: $5,000
Donor: XYZ Foundation

XYZ Foundation's investment will provide

- A $1,500 laptop computer to be used to develop curriculum and to develop educational Web site content, as well as for direct use in class slide-show presentations
- $3,500 of supplies for amphibian workshop and reptile camp

Special Notes: Please remember that I need to submit a grant report to the funder. I need to inform the funder of the impact the grant made on your students or program. So please keep track of the number of children and classes served by the grant. These funds must be spent within one year. Also, a good way to develop an ongoing relationship with the donor and maybe receive future funding is to provide the funder with photos of the children participating in a class or a letter from the kids, parents, teachers, or administrators of the community-based organizations we serve so that the funder can see his funds in action.

If you have any questions, feel free to call me at xxx or e-mail me at xxx@xxx.com.

Your name

cc: Education Director
cc: Executive Director
cc: Board President
cc: Board Treasurer

The final report shows how you can provide appealing information that reminds the grantor about the wonderful services you provide and how you touch the lives of your audiences. The cover letter gives the inside scoop on what the recipients of your services think about your program. These testimonials are proof that the funder's grant was invested wisely. To add further power, you invite the grantor to come see the program for himself.

The report communicates clearly that your organization carried out the grant as promised and again reminds the grantor of your mission and how your organization executes that mission successfully. The numbers are specific and personal.

This succinct report and letter will elevate your organization in the eyes of the grantor and will advance your relationship-building efforts. Who wouldn't want to give your organization a second grant after a sincere and effective report like this?

However, not all final reports bring great news. See Chapter 10 to review how to turn a potentially negative report into a positive one that actually bolsters your organization. By being candid and communicating your challenges in carrying out the project, how the challenges affected the final results, what the final results were, and what your organization learned, you'll demonstrate that all was not lost.

Sample Final Report

Science Camp Outreach Program Final Report

Reaching the Goal

XYZ Corporation's 2010 investment of $5,000 is very much appreciated and allowed hundreds of disadvantaged children to enjoy our unique science programs. Our grant proposal estimated that 725 children could be served with the grant award. We are quite pleased to inform you that your grant served over 1,000 children! With your support, we exceeded our goals.

Specifically, the funding was used for the following:

- 959 children (ages five years through sixth grade) participated in 30 outreach programs.
- 46 children (ages seven and eight) participated in a sleepover program with their parents.
- 1,005 total children were served.

Program Effectiveness

- Increased the number of outreach visits by 35 percent.
- 95 percent of ratings on evaluation forms received highest rating for quality of program.
- 95 percent of ratings on evaluation forms received highest rating for teacher effectiveness.

What Students Learned

The sleepover program for children and their parents was a success in creating a safe, fun family experience. The outreach visits to community organizations sparked the children's interest in the world of natural science with hands-on activities through which the children explored the mysterious world of insects, amphibians, snakes, and bats. Programs at each grade level included a demonstration, appropriate activity, and role playing to reinforce the science concepts learned.

XYZ Corporation's Impact

These programs would not have been possible without your support. Your concern for the less fortunate in the community has helped add a spark of joy and wonder to their lives! Four new community organizations have requested the Science Camp Outreach Program. Again, thank you. XYZ Corporation's continual support in the future will provide distinctive science programs to children.

Sample Final Report Cover Letter

Date

Jane Smith, Title
XYZ Corporation
Address
City, State, Zip

Dear Jane:

On behalf of Science Camp, thank you again for your generous contribution to our Outreach Program. Enclosed is our Grantee Final Report. Also enclosed are copies of thank you letters written by children who benefited from your generosity.

Here are a few comments written on evaluation forms:

"I want to thank you for the great presentation you made. Every child just loved learning about the insects. We just don't get enough of these experiences!"

"Everything was so exciting and hands on, touching the giant African millipede!"

"This has been a fantastic experience for our boys!"

"Always professional, educational, and fun—excellent program."

"The small group sizes with the fun learning program have been tremendous."

Please let me know if you or any other board member of XYZ Corporation would like to visit our site. Call me to arrange a visit to see your investment in action.

We look forward to a continuing partnership with you. If you have questions, please call me at (xxx) xxx-xxxx or e-mail me at xxxxxx@xxx.com. Again, thank you for investing in Science Camp!

Sincerely,
Your signature
Your name
Your title

14

Short Grant Proposals

The grant proposal samples that follow in the next three chapters are meant to illustrate what proposals look like so that you can see there's nothing mysterious about them. They contain the elements you've read about, and notice that they also contain the facts, figures, and heart from each organization. You have the facts and figures, as well as a passion for your organization. All you have to do is write them down as you've learned from this book, and you'll have a grant proposal ready for your readers to review.

Before you turn the page, know that the expert grant writers who generously provided their samples did so to help you and your organizations get closer to receiving grant awards so that you could continue serving the community and changing lives. However, none of the samples are meant for you to use directly; that is, do not just swap out the fictitious name and add your organization's name. It will lack your organization's personality and distinctiveness. And it will lack the personalization. You know how important it is to personalize your proposals for the grantor. Learn to write your own so that you can adapt your proposals as necessary for any grant opportunity.

Guerilla grant writers—that's you—have been trained in creating a proposal from scratch. You have the necessary tools, knowledge, and motivation to write your first proposal and many more after that. It's simple to do. The entire purpose of this book is to give you the confidence (and ammo) necessary to write as many proposals as your organization needs you to write. You can do it. You can do it in one page or two pages. If necessary, you can do it in five pages or twenty. It's still the same three steps.

I'd like to point out a few things about these samples. First, note how each organization plays on its strength. Whether that strength is—its track record, volunteer involvement, unique venue, or stellar programs—the organizations highlight those strengths in their proposals. By emphasizing the aspect that makes each organization a bit different from other grant applicants, they add potency to their proposals.

Also, you may remember that Chapter 4 mentioned that writing the perfect proposal is not the goal. If you were to go for perfection, you might never send out a proposal! With that said, in the next several pages you'll read many fantastic sample grant proposals. The samples themselves are not perfect (okay, maybe a couple of them are), but they are all effective and, most important, they were all successful.

The shortest sponsorship proposal I ever wrote was a three-paragraph e-mail. It succeeded because it was targeted to a business that was a perfect match with my organization. Through research, I found the business, confirmed the mission match, and discovered that the company had sponsored such events in the past (another set of three).

I first called the business to introduce myself and confirm its willingness to consider a new sponsorship. The proposal query was followed by a few telephone calls and a final e-mail that spelled out the terms of the sponsorship agreement. That was it. The amount awarded in both cash and in-kind items was about $1,000 to us—as well as nice incentives for our attendees. Not a lot of money, but it made our event possible. Also, by securing this lead sponsor, I was able to get a cosponsor that provided more in-kind donations for the event. This made it easier to get the "goody bag" items donated as well. Note five things as you read this proposal:

1. It mentions an affiliation with a national organization, adding to our credibility and the potential that large numbers of people would see our conference announcements (and the name of our sponsors).
2. Since the business had never heard of our organization before, I added the names of two main speakers, whom the business had heard of by the nature of its business, again adding to our credibility. The link to our organization's Web site allowed the business to see how professional we were and, what other successful activities we'd engaged in and to realize that we were a good partner that could connect it to an audience that would use its services and buy its products. The proposal informs, persuades, and assures the sponsor.
3. I asked for items the business was able to provide (the company had provided these items to other like events it had sponsored), and I aimed high, asking for a bit more, leaving the company room to negotiate.
4. My organization provided the appropriate benefits for what I was asking. (The benefits were preapproved by the leadership of the organization.). However, I left out a couple of items, leaving room for negotiation. In the end, the company got those small items as a surprise: logo recognition with a link on our Web site and a press release to industry-related magazines giving the company recognition as the lead sponsor. Throughout the months leading up to the event, I kept the sponsors informed. I e-mailed them one of the e-alerts showing

their logo recognition. I also e-mailed the sponsor the conference event program, the press release, photos from the event, and the issues of our newsletters (pre- and postconference), which gave the company more recognition. Observe that the benefits didn't cost us anything, yet they were valuable to the sponsor. (One of our members, a professional designer, designed the event program, and another sponsor donated printing services.)

5. Remember to give your lead sponsor more than the cosponsor. They should not get equal benefits; otherwise, why would the lead sponsor want to give you more cash or in-kind services if not for the added benefits? In our case, the cosponsor got a half-page ad in the event program, and all mentions of logo or name were less prominent than those of the lead sponsor.

Notice that keeping the sponsors in the loop, though just plain professional courtesy, also was a form of stewardship. My organization didn't just take the money and run. The sponsors were being appreciated and, in effect, cultivated for the next event. It also put my organization's name in front of them numerous times for a period of about six months. Do you think that these sponsors wanted to partner the following year?

The next example shows a one-page proposal. Note here the mention of the media attention the facility has attracted. The sponsor can see that this organization has the ability to provide the positive community exposure it mentions. Perhaps the organization offers a bit much in terms of benefits for the amount of the sponsorship, but this event was organized rather quickly, on confirmation of the animal's arrival from another zoo. Since the event day was near, note also a definite "reply by" date. If you offer an "exclusive" opportunity, that means that you will not seek other sponsors for that event. So use that term only if you mean it.

Sample Three-Paragraph Proposal

Hello Mary—

It was great talking to you today. As I mentioned, my group is called, XYZ Writer's Group. We are the screenwriting chapter of the XYZ Writers of America (XWA). We are hosting our first conference in San Diego on Friday, September 30. Screenwriter John Doe and Producer Jane Doe are our main speakers, and we expect about 75 attendees. We'd like to invite you to be our lead sponsor. We ask that The Business of Writing Store provide the following:

- $250 cash grant
- 80 book bags, 80 pens, and 80 bookmarks
- 80 ($10) The Business of Writing Store Gift Certificates
- Two items of your choice for door prizes

As the lead sponsor, The Business of Writing Store will receive name recognition in our e-mail blasts to XWA chapters, name recognition and a link on our Web site, and a full-page ad in our event program. If you'd like, you may set up a table to sell screenwriting-specific items such as books, CDs, DVDs, software, etc. during the duration of our one-day conference, which is 9:30 a.m. to 4:30 p.m. on September 30. You also may hang a banner from your table. We will ask other organizations to donate items to put in the book bags, such as writing magazines, highlighters, Post-It notes, etc., so we will need the bags in advance to insert the items before the start of the conference.

We're excited about this opportunity to introduce The Business of Writing Store to a new community of screenwriters and to introduce you to our members. If you have any questions, you can e-mail me or call me at xxx-xxx-xxxx. For information about us, please visit our Web site at www.xxx.org. Thank you for your consideration.

Sincerely,
Signature
Printed name
Title

Sample One-Page Proposal

Date

Name
Title
Business Name
Address
City, State, Zip

Dear Ms Smith:

We are pleased to present a unique sponsorship opportunity to XYZ Business. Your investment of $3,500 will help ABC Zoo provide a significant community event.

We are thrilled to celebrate the arrival of Jezebel, a young okapi, and newest addition to our zoo family. We are celebrating with a wild party for our zoo's supporters. Imagine the positive community exposure for XYZ Business!

With sponsorship, XYZ Business's name and logo will appear on invitations for the event. We will acknowledge XYZ Business in our newsletter and our annual report and on our Web site. You may display your banner at our main entrance on the day of the event. If you'd like, you are welcome to present the sponsorship check to ABC Zoo at the event.

ABC Zoo has been serving families of Saratoga and the surrounding communities since 1980. The zoo features 300 animals in residence and represents approximately 56 species of domestic, nondomestic, and endangered animals. ABC Zoo also offers a variety of play areas and pleasant picnic areas for families to enjoy. We offer low admission rates and free parking, which allow all families to enjoy a day in nature. Last year, 460,000 visitors experienced ABC Zoo's unique setting and our educational animal exhibits. (About 30,000 of those visitors were children who received complimentary passes because of the generosity of community sponsors).

Recently, XYZ *Magazine* readers voted ABC Zoo the "Best Park for Children," XYZ Blog rated ABC Zoo number 3 of 10 of the "Most

Popular Tourist Attractions" in Wexford County, and the *EFG Newspaper* featured ABC Zoo in an article entitled, "Educational Fun for Kids."

To take advantage of this exclusive opportunity, *please respond by July 19.* If you have any questions, please call me at (xxx) xxx-xxxx or e-mail me at xxx@xxx.com. I'll contact you in a few days to discuss this opportunity. Thank you for your consideration.

Sincerely,
Signature
Printed name
Title

The next one-page sample was written on behalf of a high school sports team. The thing to remember about writing proposals for school activities is to use the school's letterhead. A quick visit to the principal's office, and I had a couple of sheets of letterhead and envelopes. While there, I inquired if I would be able to thank the funder in the school's paper. I had, of course, talked to the coach beforehand to confirm that he wanted me to write the proposal and to get his contact information to add to the letter. Notice that I chose not to include any of my personal information in the letter. (I still could follow up with the store by saying I was calling on behalf of the coach.) The letter was successful in obtaining a gift card to the store to apply to the purchase of a new camcorder.

Note: Most funders ask for a 501(c) (3) letter, which most public schools do not have. However, the funder still may accept the proposal if the request is on the school's letterhead. Some funders may ask for a tax ID number, which the principal may have or can obtain from the school district. (Schools fall under the district's tax ID number.) Also, many public schools now have nonprofit parent organizations or foundations through which people can channel funds to get a tax deduction for the donation.

Sample One-Page Proposal

Date

Name
Title
XYZ Store
Address
City, State, Zip

Dear Ms Jones:

We would like to invite you to be a part of the upcoming baseball season at Plain Meadow High School. This letter is a request for a much-needed camcorder, new or used. We will accept a floor demo or a discontinued model. If this request is too costly, we would appreciate a gift certificate to apply toward the purchase of one. Though any amount will be helpful, $250 would truly help us to reach our goal.

This is a special year for us. The coaching staff is new, and we have plans for an exciting season. So far, parents and the community have stepped up to revitalize the facilities. Through donations of labor and materials, we were able to resurface the field, install an outfield fence, and renovate the batting cages. The school has shown support by helping to purchase vital baseball equipment.

A camcorder is essential in helping our players learn by seeing. It will be used to show each athlete how his or her individual mechanics look, and athletes' skills can be improved while increasing their confidence. Also, a camcorder will decrease injuries from using incorrect techniques. This visual process will help athletes to reach their full potential.

Please realize that any contribution is your pledge of encouragement of our student athletes, and your generous donation will improve the quality of our program. We would be pleased to acknowledge XYZ Store at our season opener and in the Plain Meadow High School newspaper. I can be reached at (xxx) xxx-xxxx, ext. 123. Thank you in advance for your support of Plain Meadow Baseball. Go Rams!

Sincerely,
Signature
John Smith
Head Baseball Coach

A sample of a two- to three-page proposal follows. It includes a cover letter that provides background information about the project. The grant proposal states directly how much is needed and how those funds will allow the requestor to reach its goals. The sample provides credible background information, addresses a pressing community problem, briefly describes foreseeable future obstacles, and offers the organization's solution. The organization's accomplishments—ensuring that all children have access to science—highlights a significant issue this country faces and at the same time reveals the organization's dedication.

Sample Short Proposal

Cover Letter

Date

Alphabet Corporation
Attn: Corporate Giving Program
123 Xxx Street
City, State, Zip

Dear Gentlepeople:

The San Francisco Bay Area thrives on challenges and innovation. Since 1996, Schmahl Science Workshops (SSW) has provided pre-K through twelfth grade children with an unmatched breadth of hands-on science workshops spanning biology, chemistry, earth science, forensics, math, and physics. The youngsters in low-income neighborhoods are just as bright as youngsters from affluent areas—they just lack the same resources and opportunities. Schmahl Science Workshops provide both. Last year, SSW provided over 700 reduced-fee workshops and put the "sizzle" into science education for 26,000 youngsters in 149 schools in San Mateo and Santa Clara counties.

In 2005, SSW served 8,542 students. Today, SSW provides exciting hands-on workshops to over 26,000 students an average of 20 times each in a year. This means over 500,000 student contact hours. The rapid growth is due in large part to our strong commitment to provide excellent hands-on science workshops to underserved youngsters.

In 2006, SSW implemented an initiative targeting underserved youngsters in low-income neighborhoods. The focus of the initiative was to provide reduced-fee workshops to schools with large numbers of Latino, Asian, and African-American students. The first year, through generous funders, SSW provided 212 reduced-fee workshops. Last year, SSW provided over 700 reduced-fee workshops. The past two years have been tough for schools. Per-pupil funding has been reduced by $1,100, and when the budget is finally adopted for the current year, more reductions are expected. There is no doubt that school districts and individual schools will have to continue reducing or eliminating many worthwhile programs, including science. At SSW, we believe in being proactive and prepared. SSW established a goal of providing 750 reduced-fee workshops this year, and we are working hard to achieve that goal. SSW is committed to continuing our efforts to provide hands-on science workshops to underserved youngsters. Therefore, we are submitting a $10,000 grant proposal for your consideration.

We look forward to developing a partnership with the Alphabet Corporation. I invite you to visit a classroom and see Schmahl Science Workshops in action. All you need is a calendar and a phone call. Developing a scientifically literate population and the next generation of scientists in Silicon Valley is too important to postpone.

Sincerely,
Signature
Printed name
Executive Director

Grant Application

Alphabet Corporate Giving Program

Organizational information:

Name: Schmahl Science Workshops, Inc.
Address:
Telephone Number:
Fax Number:
Web address: www.schmahlscience.org
Tax ID number:
Contact Name:
Title:
Telephone number:
E-mail address:

Brief description of the organization, its mission, and population served

Schmahl Science Workshops (SSW) is a nonprofit partnership of students, parents, teachers, scientists, and engineers who come together to foster the innate curiosity and love of science that exists among children. Founded in 1996 by a group of four children and their parents, Schmahl Science Workshops provides pre-K through twelfth grade children with an unmatched breadth of hands-on science workshops spanning biology, chemistry, earth science, forensics, math, and physics. Our mission is to prepare children of all backgrounds for a future in which science and technology will drive every industry and vocation.

SSW serves many schools that do not have the resources to include science as a regular part of their instructional program. SSW has an ongoing commitment to provide hands-on science to Hispanic, African-American and young women living in low-income neighborhoods. We believe that Hispanic and African-American youngsters are just as bright as youngsters from affluent communities but that they lack the same resources and opportunities. Resources and opportunities are what SSW provides every day. Evidence of the need to provide science in low-income neighborhoods is as simple as seeing how grossly underrepresented Hispanic and African-American students are at a science fair. SSW has a long-standing commitment to change the numbers of Hispanic and African-American students participating in science fairs. Seventy-two percent of the youngsters served by SSW live in low-income neighborhoods.

In 2004, Schmahl Science Workshops served 5,400 students. In 2010, SSW is in 149 schools and provides multiple workshops to 25,616 youngsters. In addition, SSW supports over 100 students conducting research in the Advanced Student Research Program. Since SSW sees the same youngsters an average of 20 times, it amounts to over 500,000 student contact hours per year. SSW puts the "sizzle" back into science!

Organization structure, including staff and board members

Like many Silicon Valley corporations, Schmahl Science Workshops started in a garage in 1996 with four students. Today, SSW is a 501(c)(3) nonprofit corporation that serves approximately 26,000 students a year with a staff of 20.

SSW's executive director provides administrative leadership. Before developing SSW, she worked as a biochemistry researcher and project manager. She is supported by a retired school district administrator specializing in curriculum, instruction, and organizational development. A retired teacher provides ongoing coaching and staff training in classroom management and workshop presentation strategies. SSW has two managers. The workshop staff is composed of instructors with strong backgrounds in one or more science disciplines. SSW is supported by 150 nationwide scientists who vet workshops as well as a small group of local volunteers involved in everything from preparing materials to fund-raising.

Currently, the board of directors is composed of five members. The composition of the board reflects ethnic, professional, and age diversity. Two board members were part of the group that founded and developed SSW. The board is working on increasing its size with additional business and community leaders. See the attached list with board member names and affiliations.

Amount requested, along with specific intent for use

If people are to be scientifically literate, science concepts cannot wait until middle school to be introduced. Not only do students need to have an interest in science nurtured at an early age, but they also need knowledge of science concepts and an understanding of the scientific process. This need is particularly acute in minority and low-income communities. The "Providing Hands-on Workshops to Underrepresented Students" project not only helps to develop scientifically literate youngsters but also provides a way to identify and encourage youngsters to become involved in science and consider science and engineering as viable career options.

Last year, Schmahl Science provided 720 reduced-fee workshops. With the current national and state budget situation, we are asking corporate and family foundations to help us provide 750 reduced-fee workshops. SSW is requesting that Alphabet Corporation provide *$10,000 for 50 reduced-fee workshops* at a cost of $200 per workshop. Since we believe that free workshops might result in a perceived diminished value of the hands-on science workshops we provide to youngsters, schools pay a nominal fee of $25.

Expected outcomes of project/program and how they will be measured

Schmahl Science Workshops has three areas that form the basis for both objectives and measurable results in the "Providing Hands-on Workshops to Underrepresented Students" project. The data collected are a part of a longitudinal study that is starting its sixth year.

- Increase the percentage of Hispanic and African-American students served by Schmahl Science Workshops from 50 to 52 percent of the total number of youngsters being served by SSW, as measured by the longitudinal demographic study and report.

- Provide 50 reduced-fee workshops to 1,500 students, as measured by reporting the number and names of workshops, the number of classes and grade levels, and the names of schools for Alphabet-funded workshops.
- Maintain or increase quality ratings of workshops by classroom teachers of at least 4.75 on a scale of 5.0 in each of seven evaluation areas, as measured by the longitudinal teacher evaluation study and report.

Conclusion

Although Schmahl Science Workshops continues to grow and thrive, we remain committed to our mission to prepare children of all backgrounds for a future in which science and technology will drive every industry and vocation. Providing quality hands-on workshops to Latino and African-American children and girls remains a core value. A $10,000 grant from the Alphabet Corporation will provide 50 reduced-fee workshops and serve 1,500 youngsters from low-income backgrounds.

In closing, at Schmahl Science Workshops we do not "read science." We do not "play science." We *do* science. I invite you or a representative from Alphabet to review Schmahl Science Workshops at our Web site, schmahlscience.org. Then visit and see Schmahl Science Workshops in action. All that is needed is a phone call. My direct line is (xxx) xxx-xxxx. Having scientifically literate citizens and developing the next generation of scientists is too important to delay. Let's see if we can work together.

15

Medium Grant Proposals

The following grant application is quite brief. (Do you see how this application would take you only a few minutes to complete if you had previously gathered those attachments?) Because this application was so short, the grant writer added a compelling cover letter that allowed her the space to personalize the proposal and add further eloquent and persuasive information. Notice in Section II, item 3, that there is a potentially negative question about the prior year's deficits. However, the grant writer deftly handles her response and also offers the organization's treasurer's contact information if the foundation wishes to inquire further. (Of course, she gave the treasurer a "heads up"!)

The cover letter was submitted on the organization's letterhead and was signed by both the board president and the board's grant writer. The cover letter also included the children's librarian's contact information in case the grantor had questions regarding the programs. You do not want to inundate staff with potential calls, so use this idea sparingly and only after preparing staff for the call. As with all the grant samples in this book, the proposal was successful in obtaining a grant award, and the names and addresses were removed for the purposes of this book.

Sample Medium Grant Proposal

Cover Letter

Date

Name
Title
Address
City, State, Zip

Dear Ms Garcia:

Thank you for giving Friends of XYZ Public Library the opportunity to submit the attached grant application for the funding of XYZ Library's Children's Programs 2011. Perhaps Ben Franklin knew when he chartered the first circulating public library in 1732 that the library would become one of the country's most revered and lasting public institutions—an institution that would promote and enable lifelong learning, personal and professional growth, and life satisfaction for Americans of every age and walk of life.

At XYZ Library, learning is indeed lifelong. In the children's library, learning starts at birth, with programs for bookworms ages 0 to 5 years. Middle school children serve on an advisory committee. While munching on pizzas, drinks, and healthy snacks, they offer ideas for programs, scheduling, outreach, decorating, and book collection—ideas that have resulted in a doubling of participation in the children's programs and leadership experience for them. Tweens and teens are active in the summer reading program and also take leadership training to assist in the children's library, reading stories and helping with homework. One day they may return with their own offspring to start the cycle anew, remembering that they were once a part of one of the valley's most innovative and well-attended children's library programs.

Membership dues, used book sales, and grants enable the Friends of XYZ Library to support many enriching, innovative programs—programs that introduce preschoolers to books, computers, and the love of learning at an early age and keep them involved through the teen years. For the past decade, Friends have provided funding for 350 children's library programs per year, attended by over 12,500 youngsters from birth to teens. The Friends have provided money for computers, software, and programs that cannot be funded through the library's operating budget.

With nearly 41,000 children's items available for checkout and approximately 30 programs hosted each month, there is something for every age group at XYZ Library, and helping to make the library a friendly, inviting place for children and teens is an important goal of Friends' fundraising. We hope that ABC Foundation will provide 2011 funding that will encourage kids to become lifelong readers and learners and give parents the reassurance that the library is a safe, comfortable, and good place for children and teens.

Attached please find the application with attachments. Thank you for your consideration of this request. If you have questions, please contact Mary Smith, Friends' grant writer, at (xxx) xxx-xxxx, or Jane Jones, XYZ Library, Children's Librarian at (xxx) xxx-xxxx.

Sincerely,

Signature

President, Friends of XYZ Library

Signature

Grant writer, Friends of XYZ Library

Grant Proposal

2010 Grant Application

Instructions

Please be sure to review the Grant Guidelines before completing this application. Please keep your answers as brief as possible.

All Grant Applicants: Please complete all of Section I, include the required attachments, and sign and date the application.

Section I

Name of Organization: Friends of XYZ Public Library
Federal Tax Identification Number: ______________________
Address: ______________________
City: ______________ **State:** ________ **Zip code:** __________
Telephone: ______________ **Fax:** __________
Web address: ______________________
Executive Director: ______________________
Telephone/Fax: ______________________
E-mail address: ______________________
Secondary contact: ______________________
Telephone/Fax: ______________________
E-mail address: ______________________

1. **Amount requested:** $12,000. Date of application: September 1, 2010.
2. **Type of request** (check one): ☐ Operating ☐ Capital ☒ Program ☐ Project
3. If the request is not for operating support, briefly describe the program or project for which the organization seeks support.

Program: Year-Round Children's Library Programs: Funding for 2011

Friends of XYZ Public Library provide funding for 350 children's programs per year attended by over 12,500 preschoolers, school-age children, and teens. Friends' membership, used book sales, and grants enable many enriching programs that introduce preschoolers to books and the love of learning at an early age and keep them involved through the teen years.

With nearly 41,000 children's items available for check out and approximately 30 programs hosted each month, there is something for every age group. School-age children enjoy year-round book clubs and programs such as bicycle, skateboard, and pedestrian safety; Spanish story time, origami from the ABC High School Japanese Honors Society; and meeting guide dogs and puppies-in-training through Canine Companions for Independence. Emmy Award–winning storyteller Diane Ferlatte enchanted the children with her "Dream of King" stories, honoring the life of Martin Luther King, Jr. Volunteers donate their time to lead Spanish story time, teach knitting, assist with story-related crafts, and help with homework.

This year 30 middle school students developed leadership skills while participating in the library's new Middle School Advisory Group. The young advisors discussed their ideas for library programs, scheduling, outreach, decorating, and book collection while snacking on pizzas and beverages funded by Friends of the Library. The group's ideas have resulted in popular new programs and projects that have produced an increase in participation. This spring they helped library staff plan the first ever Middle School Summer Reading Program that attracted 75 participants, over 60 percent of whom completed the program. Based on the group's success, the library plans to develop a teen advisory group.

4. Does the request address one of our funding priority areas? ☒ Yes ☐ No.
   If yes, select one: Cultural and Artistic Enrichment
5. Has the organization received a grant from our Foundation in the last three years? ☐ Yes ☒ No. If yes, please list dates and amounts.
6. Please list any of our employees involved in your organization and their roles.
   N/A____

   ____
7. Does the organization receive support from United Way? ☐ Yes ☒ No

8. Please provide a brief overview of the organization:

Friends of XYZ Public Library is a nonprofit organization established to encourage use of the library for the pursuit of knowledge, information, and pleasure for persons of all ages and walks of life; increase public awareness of library resources; raise funds for library enhancements; and to provide volunteer help to the library and its patrons.

9. Of the clients you serve, what percent are in the following categories? N/A

XYZ Public Library does not track or limit access to its free, life-enhancing books, services, and programs. The free public service is available to persons of all incomes, ages, socioeconomic status, and ethnic groups.

Income
Do not track. Library is open to all without charge or eligibility requirements. ☒

Ethnic groups
Do not track. Library is open to all without charge or eligibility requirements. ☒

Gender
Do not track. Library is open to all without charge or eligibility requirements. ☒

STOP: If the request is less than $5,000, skip Section II. Please sign/date the application and include the required attachments.

Section II

Financial Information

1. The organization's current year budgeted expenses of $103,589 are 45% ☒ higher ☐ lower than the previous year's actual expenses.
2. During the current fiscal year $60,240 or 58 % of the total expense budget is for administrative/overhead and fundraising expenses.

3. Has the organization experienced an operating deficit (i.e., expenses exceeded revenues) in the last two years? ☒ Yes ☐ No. If yes, what was the amount of the deficit?

June Year 2009	Deficit $31,558
June Year 2008	Deficit None

4. Please explain the deficit(s) above and the plan for reducing or eliminating it.

Capital campaign startup expenses for funding the new library's furniture and fixtures. Startup expenses include our campaign consultants and hiring a development director. The deficit will be eliminated as the funds come in over a period of about one year. For questions related to the Friends budget, please contact Treasurer John Doe at (xxx) xxx-xxxx.

Project Information (complete only for program, project, or capital support)

1. What are the timelines for the project and for fundraising?
   This grant request is for funding of the Library Children's Programs for January 1 through December 30, 2011
2. The budget for the program? $ 15,500 = $12,000 grant request $3,500 from used books sales
3. How does this effort address a community need? Please describe the community and clients that will benefit.

A well-educated, well-informed citizenry begins with childhood exposure to books, learning, and cultural programs that are available at our public library and with encouraging youth to regularly visit one of the oldest and most revered institutions in America. The library is a safe place for children and teens to learn, interact, and develop leadership skills through exposure to the printed and electronic word and special programs.

4. Please explain how you have measured or will measure the success of the program/project.

Project and program successes are measured by the number of participants and feedback from children and parents.

Section III. Required Attachments for All Grant Applicants

Please enclose one copy of each of the following items:

☐ Cover letter
☐ A copy of your current Internal Revenue Service (IRS) determination letter indicating tax exempt 501(c)(3) status
☐ Board of directors list, including names, phone numbers, and affiliations
☐ Annual report, if available, or other material summarizing activities of the organization
☐ Current year itemized operating revenue and expense budget for the organization
☐ Most recent audited financial statements or IRS Form 990
☐ A list of major corporate and foundation donors for the past two years

If you completed Section II, please also enclose one copy of each of the following items:

☐ A one-page summary of the organization's three major core programs or activities
☐ Budget of program, project, or capital campaign
☒ Please find Items attached

Authorization

The undersigned certifies that they are authorized to represent the organization applying for a grant and that the information contained in this application is accurate. The undersigned agrees that if a grant is awarded to the organization:

(*1*)The grant will be used for the purpose outlined in the grant award letter and may not be expended for any other purpose without prior written approval from our Foundation, and
(2)information about the organization and the grant may be used by our Foundation in any published materials.

____________________________________ _______________

Signature of Executive Director or Board Chair Date

The next medium-length grant proposal is different from the others we've discussed. This one is for the preservation of animals, specifically, orangutans. Note, however, that it is no less compelling. It clearly reveals the dedication of the organization applying for the grant and the people they serve so that together they can preserve critically endangered orangutans.

This proposal shows, in four pages with a one-page project budget, the history of the organization, the incredibly complex nature of the project—spanning multiple countries—the people involved, past successes, obstacles they face, and exactly how the project will help them overcome the obstacles. The structure of the proposal (with nice headings) leads readers along, providing relevant information, so that even if they had never heard of the organization before, they would get the sense that the organization had the experience, connections, and commitment to make the project a success. The proposal also included other relevant attachments, which were left out of this book. Targeted to funders that are just as passionate about protecting endangered animals as this organization, how could it not succeed?

Sample Medium Grant Proposal

Orangutan Conservancy (OC) 2010 Veterinary Workshop

Abstract

The Orangutan Conservancy (OC) is seeking $3,500 from the XYZ Foundation to help stage the OC 2010 Veterinary Workshop, designed to bring together veterinarians and wildlife health experts from Borneo and Sumatra devoted to issues of captive care. This workshop, which will be staged August 2 through 6 in Sumatra, will build on the success established by the inaugural workshop held in 2009, where 35 delegates from Indonesia and Malaysia gathered to hone their skills, share their knowledge, and build the partnerships necessary to rescue orangutans successfully.

The OC 2010 Veterinary Workshop also will include veterinary personnel from gibbon rehabilitation centers in Southeast Asia, in addition to local, regional, and national veterinary workers. The workshop will engage the staff members who provide daily veterinary care and will cover issues such as crisis care, nutrition, emerging infectious diseases, and contraception to help extend the network of mutual support among the delegates.

Introduction

Orangutans are in severe crisis. The largest of the great apes found in Asia, their natural range is limited to the islands of Borneo and Sumatra, and their rainforest homes are disappearing quickly. More than 80 percent of the orangutans' habitat has been destroyed over the last 20 years, and no more than 63,000 orangutans are thought to exist. At the current rate of decline, experts believe that orangutans may become extinct in the wild within 25 years.

The primary threats to orangutans are illegal logging and habitat destruction, human encroachment, the conversion of rain forests to oil palm plantations, and the pet trade. As a result of such intense pressures, an extremely large number of orphaned orangutans exist in rehabilitation centers across Borneo and Sumatra. These orangutans—which number approximately 1,600—arrive bearing a host of physical and emotional wounds and require intense veterinary care to recover.

The orangutans that are judged fit to return to the wild are reintroduced through a long, complex process, but the overwhelming majority continues to reside in the rehabilitation centers. Unfortunately, orangutan rescue, rehabilitation, and conservation efforts are compromised by a difficult set of obstacles, including political opposition, official corruption, and marketplace pressures. Equally confounding is the traditional lack of coordination and cooperation among the rehabilitation centers in Indonesia and Borneo, which often exist in direct competition with one another.

The Orangutan Conservancy (OC) is seeking $3,500 from the XYZ Foundation to help stage the OC 2010 Veterinary Workshop, designed to bring together veterinarians and wildlife health experts from Borneo and Sumatra from August 2 through 6 in Sumatra for a seminar devoted to issues of captive care.

Although designed primarily for the veterinarians and health care workers at orangutan and gibbon rescue centers, the OC 2010 Veterinary Workshop also will be open to local, regional, and national veterinary workers in both Borneo and Sumatra.

The inaugural OC Veterinary Workshop that was staged in 2009 was the first conference ever designed to specifically address the lack of communication and coordination between the rescue and rehabilitation centers. By focusing on the veterinary staffs—rather than facility managers or founders—the OC was able to sidestep the traditional politics that have hindered conservation efforts in the region and begin to empower a generation of passionate, skilled veterinary health care workers in the process.

Goals and Objectives

The OC 2010 Veterinary Workshop will focus on the latest health and veterinary issues among primates, both in captivity in rehabilitation centers and in the wild, and ensure that each facility works from a baseline of minimum standards and best practices. Delegates to the inaugural workshop in 2009 have already begun preparing the agenda, fueled by the knowledge that the current cycle of confiscation and long-term care cannot continue without an ulterior goal. The OC hopes that by refining the veterinary care, encouraging a network of mutual respect and support, formalizing the protocol, and creating national participation, we can reverse the growing trends toward extinction.

The OC 2009 Veterinary Workshop was the first of its kind in Indonesia or Malaysia, a gathering specifically targeted not at the leaders or visionaries who founded the rescue and rehabilitation centers but rather at the national staff members who play no less an important role in the orangutans' survival. It served to break down many of the barriers—if only for a week—that exist between centers and set the stage for a new era of communication and respect. The OC believes that a workshop such as this can lay an important foundation for future workshops and future conservation efforts.

The OC 2010 Veterinary Workshop will focus on three main objectives:

- Raise the level of care and welfare for orphaned orangutans at rehabilitation centers:
    - Define minimum standards.
    - Identify best practices.
    - Teach basic skills.
- Create a community of support and trust among rehabilitation center staffs:
    - Encourage sharing of ideas.
    - Develop mutual respect.
    - Exchange support.
    - Present case studies that promote discussion.
- Raise the capacity of national staffs:
    - Import training and facilitation from the United Kingdom, Australia, and the United States.
    - Encourage accountability.
    - Establish a foundation for future exchanges.

In addition to those main areas, the OC 2010 Veterinary Workshop will touch on other topics as well, such as (1) education, (2) conservation, and (3) animal welfare.

Education

Although community sensitization and conservation outreach are important aspects of primate conservation in Asia, it is equally important to educate the staffs at the rescue and rehabilitation centers in Borneo and Sumatra. The lack of capacity training and authority afforded the national staffs is traditionally very low at these facilities, even though the ultimate responsibility for the running and management of the centers will one day be theirs. The Orangutan Conservancy is mindful of the obstacles to such empowerment—including language skills, cultural sensitivity, mutual distrust between rehabilitation centers, and outdated management practices, among others—but believes that this process is long overdue. By giving these veterinary health care workers the tools and inspiration to improve, the OC believes that positive change can be effected.

Conservation

Ultimately, the conservation outcome will be the successful reintroduction of endangered orangutans back into the wild, a process that will take years to complete. But short-term conservation benefits also must be considered, including (1) increasing the capacity and training of Indonesian and Malaysian veterinarians and health care workers at both rehabilitation centers and in surrounding wildlife offices, national parks, and universities; (2) raising awareness within Indonesia of the importance of orangutan and primate health issues; and (3) presenting and discussing case studies from each rescue center that might head off a future disease or health crisis. However, the Orangutan Conservancy 2010 Veterinary Workshop will maintain as its ultimate conservation goal the creation of a network of respect and trust that can play a pivotal role in the rescue and rehabilitation of orangutans.

Animal Welfare

The Orangutan Conservancy and the projects it supports in Malaysia and Indonesia are dedicated to the protection of orangutans and their rainforest homes. As a result, animal welfare is paramount, as is a determination to provide the best possible veterinary care. Although there could be a practical session that is performed on an orangutan at the host facility, it will only be a procedure that is necessary and beneficial to the long-term health of the animal. At no time would the Orangutan Conservancy or any of the organizations associated with this workshop seek to needlessly harm, anesthetize, or in any way compromise a captive animal.

Methods

The OC 2010 Veterinary Workshop will follow the basic format of presentations, case studies, and practical demonstrations that will allow delegates to be exposed to a wide variety of husbandry issues. Three of the five days will be held in a conference-style format, with presentations and working groups designed to familiarize delegates with existing cases and conditions at orangutan rehabilitation centers across Indonesia and Malaysia. The other two days will be spent on practical sessions at the Sumatra Orangutan Conservation Project (SOCP), where delegates will get a chance to implement skills and techniques they have learned on animals in need of medical treatment.

At the conclusion, a workshop report will be distributed to each of the delegates, both in printed form and on CD, along with a CD of the presentations. As noted earlier, the agenda itself will be based on the results of the questionnaire distributed to orangutan rescue centers.

Project Impact

The OC 2010 Veterinary Workshop will follow the same format that was introduced at the inaugural workshop a year ago, one that has already had an impact on the style, skills, and resources associated with veterinary care. Before leaving the 2009 workshop, the delegates created the Orangutan Veterinary Advisory Group, their own network of advice and consult, and established an orangutan veterinary list-serve specifically designed to post and comment on case histories at their facilities.

The veterinarians also created a formal request for funding to establish a nationwide microchip program in order to standardize record-keeping among the centers and promote faster communication and greater transparency.

The orangutan veterinarians also have committed to produce their own Crisis Care Manual based on the belief that no existing textbook addresses the injuries and illnesses in orangutans that they face on a daily basis. A first draft of the manual should be available for review at the OC 2010 Veterinary Workshop.

Perhaps the best gauge of the impact of the OC Veterinary Workshop in 2009 came from the delegates themselves. Said a delegate from Borneo: "This was a dream come true. … We have gained confidence. Getting the feeling that we are not alone, that we have friends who are willing to help us is so nice. Also, the feeling of being 'united' gives us confidence when we have to talk to the government."

Timeline

The OC 2010 Veterinary Workshop will begin on August 2 and conclude on August 6. Within 60 days of the conclusion, an OC 2010 Veterinary Workshop Report will be distributed in either electronic or printed form to all delegates, sponsors, and interested individuals. But the timeline for this workshop actually began more than a month ago, when delegates from the inaugural workshop concluded that seminar by meeting to lay out the goals and objectives of the 2010 workshop.

OC 2010 Workshop Budget

Item	Unit Cost	Total Cost
International airfare	4 @ \$2,500 per	\$10,000
Regional airfare	26 @ \$200 per	\$5,200
Accommodations	30 @ \$55 per × 6	\$9,900
Ground transportation	NA	\$1,000
Printed materials	NA	\$500
TOTAL		**\$26,600**

Funding Notes

- Funding requested from the XYZ Foundation will be used to offset travel costs of the Indonesian, Malay, and other Asian delegates.
- The Association of Zookeepers has committed \$15,000 to the project.
- The Orangutan Conservancy has committed \$3,000 to the project.

The next organization called its sample proposal a combination grant and sponsorship request. It differs from other grant proposals in that it offers benefits to the funder, similar to a sponsorship request, yet the request was for a cash grant. The XYZ Nonprofit Organization's proposal describes the impact on children of not having the necessary school supplies. Something as simple as a backpack, an item often taken for granted, is convincingly conveyed as a crucial ingredient to children's preparation for school and their success in learning. The organization describes the enormous volunteer base involved (community support of the need) to fill and distribute the backpacks to the targeted children.

Sample Medium Grant Proposal

XYZ Nonprofit Organization School Drive
Grant/Sponsorship Request

May 19, 2010

Contact Information:

Organization Name
Street Address
City, State Zip
Office: (XXX) XXX-XXXX
Fax: (XXX) XXX-XXXX
Contact: Name
(XXX) XXX-XXXX

Organization description and mission

The XYZ Nonprofit Organization was created by our two founders in 1990 as a university MBA class project with the intent of making a positive and lasting impact on children in need in the community. This intent has become our mission.

Our executive director and cofounder recognized a need to support local underprivileged students and created the school drive in 2001. Now in our *tenth year,* this program has provided new backpacks and school supplies to more than 91,028 underprivileged children in the greater Bay Area and Silicon Valley. This year the program is collaborating with approximately 260 corporate partners to serve more than 130 schools and social service agencies. In 2009, the program provided 13,843 students with new backpacks and supplies.

The XYZ Nonprofit Organization is also celebrating its twentieth year of serving children! The XYZ Nonprofit Organization began in 1990 with the creation of the Gift Drive, which provides specifically wished-for holiday gifts to underprivileged children. Our founders provided more than 2,000 gifts to children in East Palo Alto that year. In 2009, the XYZ Nonprofit Organization provided gifts to 61,068 low-income children in partnership with 244 social service agencies and 935 corporate partners.

Program description

The XYZ Nonprofit Organization School Drive provides new, age-appropriate backpacks and school supplies to low-income children living in Santa Clara, San Mateo, San Francisco, and Alameda counties. *All* the children we serve are enrolled in the federal Title I Free or Reduced Price Meal Program. *To qualify for this program, a family of four, for instance, would have a total income at or below* $28,665. After basic needs are met, such families seldom have money for school supplies.

The success of the School Drive is based on numerous important partnerships established with schools and social service agencies (where children and their families are registered) and locally based corporations that provide financial support and backpack donations. Partner schools and agencies register with the XYZ Nonprofit Organization each June by filling out backpack request lists for the children they serve. Requests are then distributed and fulfilled by our corporate partners and their employees.

Geographic areas served

The XYZ Nonprofit Organization anticipates serving approximately 15,000 underprivileged Title I school children in Santa Clara, San Mateo, San Francisco, and Alameda counties in 2009. Of the children we help, 69 percent live in Santa Clara County, 16 percent live in San Mateo County, and the remaining 15 percent live in other Bay Area counties, primarily Alameda and San Francisco counties.

Informing the target population

This year we are partnering with over 120 Title I schools and 13 human service agencies. The partner schools can easily identify which children are in the Title I program because they are qualified by federal standards. Likewise, our agency partners have eligibility levels at or below the federal poverty level. Having each child prequalified for need through the schools and agencies ensures that every child is truly a child in need.

Volunteers

We will have approximately 675 to 725 volunteers work in our donated warehouse in August. Our volunteers come from corporations, foundations, rotary clubs, Boy and Girl Scouts, community organizations, families, and individuals. Our volunteers are essential to the low-cost operation of our program. Many come back year after year.

Part of our mission is to introduce philanthropy to our communities. Our warehouse accommodates families, young students, and many corporate volunteers who are participating for the first time. We believe that this exposure will benefit our community, no matter where such people choose to spend their volunteer time.

Measurable program objectives

- Increase volunteers from 650 (2009) to 700 (7 percent).
- Increase host companies from 285 (2009) to 300 (5 percent).
- Distribute at least 15,000 supply-filled backpacks, donations permitting.
- Partner with at least 130 schools and social service agencies.

Measuring success/evaluation plan

At the end of our program in early September, we hold several wrap-up sessions with staff, during which each aspect of the program is evaluated. These meetings are in person and include several of our key volunteers, corporate hosts, and warehouse co-coordinators. Recommended changes are implemented the following year to increase efficiency and reduce cost.

We gather information through e-mail and phone surveys, which are conducted in September to solicit valuable information and suggestions from volunteers, corporate partners, and agencies. Their responses are reviewed and incorporated in the following year's program planning. These evaluation tools have led to many improvements throughout the years.

Description of past success

The School Drive has seen a steady increase in the students it has been able to serve, beginning with 3,200 in 2001. Since then, the program has continued to help more Title I students each year. All told, the School Drive has provided new backpacks and school supplies to more than 91,028 students since 2001.

Foundation Acknowledgment

For a complete description of sponsorship levels and benefits, go to our web site at http:xxxxx.com Following is an example of some sponsor benefits:

1. Acknowledged as a major School Drive funder in the San Jose newspaper with an advertisement following the end of the program
2. Recognition in the XYZ Nonprofit Organization monthly newsletter (readership of over 19,000)
3. Recognized as a major funder at our annual celebration luncheon in September
4. Prominently recognized as a major funder on the XYZ Nonprofit Organization Web site

Recognition/awards

For the past 3 years, Charity Navigator awarded the XYZ Nonprofit Organization a four-star rating, the highest possible, for our efficiency and fiscal responsibility (www.charitynavigator.org).

Our executive director was chosen as the "2006 Business Person of the Year" by the Chamber of Commerce and was selected as the "Distinguished Alumna of the Year" by the X University's College of Business. In 2007, she won the "Woman of Distinction Award" given by the *Business Journal* and the National Association of Women Business Owners. In addition, we have received numerous recognition letters (copies available on request).

2010 Goals

During the process of surveying the Title I schools in Santa Clara and San Mateo counties, we learned that there are 15 elementary schools in these two counties where 100 percent of the students are eligible for the free or reduced-price meal program. These particular schools are struggling to teach 7,161 students with very little resources. Massive budget cuts leave them unable to hand out supplies, and most of the students arrive on the first day with little or nothing.

These schools, for the most part, have no active PTA support. They are in communities where both parents work to make ends meet, and fundraisers are ineffective. The San Jose newspaper recently ran an article on fundraising to save programs and teachers and provide supplies. The XYZ Elementary School District, where most of the 100 percent schools are located, held a jazz festival that earned only $1,450. In comparison, the ABC School District held an online fundraiser that earned over $2 million. This allowed the district to cancel all teacher layoffs and restore cut programs.

This inequity is very disturbing to us, and we are determined to provide supply-filled backpacks to all these 100 percent schools.

We believe that one of the greatest things we can provide a child is hope and a future. Education is a critical pathway to a life beyond the cycle of poverty and destructive lifestyles. Something as simple as a backpack and school supplies shows a disadvantaged child that he or she matters and that people care about him or her. In addition, a backpack can engender enthusiasm and excitement for school in the hearts of children who before may have felt only hopelessness in their situation.

Will you join us in reaching these children? A pledge of support will be used to bulk purchase new high-quality backpacks and essential school supplies at a deeply discounted rate. A $10,000 grant will provide approximately 670 supply-filled backpacks, enough for every student in two of the 100 percent schools we will serve this August. Help us send these children to school prepared to learn and succeed.

We are so grateful for your generous support of the School Drive in 2009. Thank you for your consideration again this year.

Signature Development Director
Organization Name
Street Address
City, State, Zip
XXX XXX-XXXX. ext. XXX
E-mail address

The last medium proposal, from the Environmental Volunteers, offers an example of a grant proposal that requests renewed funding for program development. (Renewed funding means that the funder supported this same project previously.) Note that the organization's background is nicely articulated in one paragraph. This proposal describes the organization's core programs that the grant would support and specifically informs the grantor of the program's goals and objectives. The numbers involved and the commitment of the schools to participate over multiple years evidence the wide-ranging community support of the program. All are important factors when requesting funds to support such a program.

Sample Medium Grant Proposal

The Environmental Volunteers
Proposal to the XYZ Foundation

Mission

To promote understanding of and responsibility for the environment through hands-on science education.

Background

In 1972, when awareness of the threats to the San Francisco Bay ecosystem began to grow, a group of citizens concerned about preserving the estuary for future generations created a small, hands-on program and recruited volunteers to take local schoolchildren to the Bay so that they could experience it, know it, enjoy it, and grow to take care of it. This was the birth of the Environmental Volunteers (EV). Now, 38 years later, our programs continue to engage children in respectful environmental stewardship. Using innovative hands-on learning stations and inquiry-based teaching, volunteer docents help children to develop an awareness of today's pressing environmental issues while inspiring a sense of stewardship to help preserve our natural heritage. Our university-accredited training program prepares hundreds of volunteers to deliver environmental science education through classroom lessons and activities, field study, weekend excursions, and summer science camp programs. Annually, our classroom-based science education and field trip programs now reach more than 12,000 students and their 400 teachers. Our summer camps allow over 100 youth, many from low-income communities, to spend a week exploring the wonders of the natural world at local natural sites. For many children, participation in our programs sparks their first awakening to the wonders of science, the natural world, and environmental stewardship.

Core Programs

Our signature program, *Kids-In-Nature Environmental Science Program,* includes classroom and field study activities for K through eighth graders in nine subject areas: Baylands Ecology; Forest and Foothill Ecology; Earthquake Geology; Marine Ecology; All About Birds; Early California Indians: An Environmental Perspective; Water Science and Conservation; Nature in Your Neighborhood; and Energy and Natural Resources.

The natural world is an exciting and inspiring place for children to develop skills in analytical thinking and scientific methods. Central to the EV methodology is an inquiry-based approach to teaching. This involves the use of hands-on learning kits that give children the opportunity to explore while developing a keen awareness of and enthusiastic appreciation for our shared environment.

The EV has 140 unique learning kits that teams of volunteers set up as discovery stations to inspire, enthrall, and capture the imaginations of students. Because all programs are linked to state science standards, teachers receive valuable assistance in delivering their curriculum.

Through our *Equity and Ecology Program,* approximately 25 percent of our science programs are delivered to low-income classrooms. These schools receive our services free of charge or at substantially reduced rates.

The EV also conducts *summer camps* for both elementary and middle school students. Common to all our programs is an emphasis on the excitement of learning, the importance of science, and the role each of us can play in being good stewards of the environment, as modeled by our engaging, caring, and environmentally committed docents. Key to our successful management of over 100 teaching docents is our university-accredited *Volunteer Docent Training Program,* which serves our own volunteers, as well as park rangers, docents from other organizations, and school volunteers who turn to us for training.

Grant Requested

We respectfully request a grant of $40,000 to support the continued development and success of our *Educate for Depth* and *Science by Nature* initiatives.

Grant Purpose

In the natural science education sector, it is common to offer single programs to as many individual classrooms as possible, scheduling at most one classroom service followed by a single field trip. This delivery model reaches as many children and teachers as possible, but it results in very few hours of instruction per child. *We set out to address the challenge: Can we deliver more hours of instruction to classrooms and produce measurable academic improvement in a resource-efficient manner?*

In 2004, we designed a three-year pilot program in conjunction with a local low-income school. To further advance students' academic performance, the pilot program tested a new approach to service delivery. We built a "continuum of learning" by linking the existing programs of multiple peer organizations and focusing those programs on the same classrooms. In the pilot program, six organizations collaborated and delivered multiple classroom programs and field trips and ensured that all programs were the foundation for or reinforcement of the other programs. Through this innovative approach to service delivery, students received at least 40 hours of science instruction annually for three years with multiple field study excursions. An independent evaluation study found significant academic improvement (average increase on posttest was 76 percent) and a resulting strong commitment to environmental actions.

Based on the success of the pilot program, we then set out to scale this model so as to reach significantly more students. With the support of the XYZ Foundation, we launched two initiatives last year to bring this program to scale. We request a renewal of that partnership and support to continue the successes we are achieving with these initiatives.

1. *Science by Nature* (www.sciencebynature.org). This Web-based system enables large numbers of teachers to select environmental science programs delivered by multiple independent organizations. The Web system guides teachers to programs that meet curricular needs and suggests sequencing of programs for the greatest educational impact. Teachers use

one efficient, common application form to register for programs from multiple providers. Collaboration partners coordinate the delivery schedule of the requested programs so that the programs provide reinforcing education. All programs are searchable by grade, subject, and state standards that can be met. We launched the Web site in October, initially with the program information from four collaboration partners. There are an additional eight organizations in the collaboration, and their program data will be added over the coming months. We expect to add new partners to the collaboration as well. In the first week following the launch, 25 teachers used the site, and they requested a total of 53 programs.

2. *Educate for Depth*. Because a majority of our programs are *not* delivered in collaboration with other organizations, we set out to change our own delivery model so as to achieve greater educational depth. We now preferentially serve schools that commit to environmental science programs for multiple consecutive grades over consecutive years. For example, by serving the third and fourth grades consecutively, students receive reinforcing programs over two years. There has been a very strong response from schools; some have even committed to programs for every class and every grade (a five-year continuum of learning!).

Through these two initiatives, we seek to

- Strengthen the collaborative efforts of regional environmental peer organizations and link existing programs, thereby resulting in more hours of education being delivered in a resource efficient manner;
- Improve and expand science learning at the K through grade five level in San Mateo and Santa Clara counties; and
- Create a student population that is measurably more proficient in science, more environmentally conscious, and more academically enriched.

To achieve these objectives, we will

1. Serve entire grade populations within participating schools with multiple, connected science programs delivered over multiple, consecutive years.

2. Foster collaborations with peer education organizations to "educate for depth" and link the programs of multiple organizations in a logical fashion so as to collectively provide greater resources to participating schools.
3. Scale the Science by Nature Web-based system to facilitate participation by a large number of teachers and multiple science education organizations.

By June 2010, the Educate for Depth program will achieve the following:

- Collaborations at two low-income schools will be renewed and programs delivered successfully by all collaboration partners.
- Programs will be delivered at 19 schools that have committed to at least two years of natural science education programs for all classrooms in two or more consecutive grades.
- All current participating schools will be approached to renew their commitment for the subsequent school year, and at least 80 percent will renew.
- Metrics will be established for program growth in the 2010–2011 school year, and a plan will be defined to achieve that growth.

By June 2010, the Science by Nature program will achieve the following:

- Collaboration agreements will be renewed with at least 80 percent of current collaboration partners.
- A plan will be defined for outreach to additional new partner organizations.
- At least 70 teachers will use the Web-based system to request programs from at least two collaboration partner organizations.

Organizational Budget:

FY 09–10 budget: $541,001

Current staffing:

- Professional staff of seven (five FTEs)
- Four interns (local college students in education or environmental studies who deliver classroom programs, thus allowing us to increase capacity)

Project budget: $203,769

Budget attached

Accomplishments since the last funding (if requesting a renewal)

We submitted our Final Report very recently. Accomplishments (and challenges) referenced in that report represent the current status quo.

Board List

See attached.

We have 100 percent board participation in our annual campaign.

16

Long Grant Proposals

The next five proposals demonstrate effective long grant proposals. The longer length is used to go into depth about the organization, the project, or any other issues about which the funder requires further details. On the other hand, some funders ask several questions in their grant applications, and you have no choice but to answer all of them, taking many pages to do so. (Look at the variety of questions asked and answered in these samples.) Still, guerilla grant writers are trained to describe succinctly and communicate relevant and convincing information while inspiring the funder to award them a grant. The longer page count does not give you permission to ramble!

These five sample proposals, as well as all the others in this book, are uniquely formatted, showing that you have some freedom in presenting your proposal. As long as you are careful to stay within the word limits or page limits of the application (if any), you should use whatever structure makes it easy for readers to follow along and find the answers to their questions.

The Hospice of the Valley proposal requests two-year support for a new project. Notice the specific description of how the funds will be used? The sample from the Children's Discovery Museum of San Jose requests renewed funding for program support that covers staff salaries. The proposal provides an in-depth and impressive description of the program. The Bay Area Glass Institute's proposal is a dynamic example of an arts organization accentuating its one-of-a-kind status in the entire county. The proposal for the Homeless Agency for Intact Families requests funds as part of a capital campaign. Note the time taken to discuss the campaign and the partner involvement in the campaign. The Best Aquarium in the Whole Wide World's proposal illuminates the world of aquariums and provides significant substantiation for renewed general operating support.

In all the proposals, note the compassion and heart as well as the powerful use of personalized numbers. These sample proposals also convey the organizations' capacity to successfully carry out the projects. Their remarkable track records encourage a sense of trust.

Sample Long Grant Proposal

Application for Funding

Name or Organization: Hospice of the Valley
Date of Application: 7/15/10
Address: __________ **City:** __________ **Zip:** __________
Contact: __________ **Title:** Director of Development
Phone: __________ **Fax:** __________ **E-mail:** __________
Tax ID: __________
Date/Place of Incorporation: __________

Please briefly state the mission of your organization and its long-term strategic goals

Hospice of the Valley (HOV) was the first nonprofit hospice care program in Santa Clara County, and this year marks our thirtieth anniversary of community service. Since 1979, HOV has served 30,000 individuals and their families facing life-limiting illness at the end of life. Our mission is to affirm dignity, hope, and comfort for those facing a life-limiting illness by providing compassionate palliative, hospice, and grief care. Our professionals and volunteers serve the physical, emotional, and spiritual needs of individuals and their families. We enhance our community through education, research, advocacy, and volunteerism.

All major programming decisions are made in alignment with our current Strategic Plan goals, which are as follows:

- *Quality in service.* We will be committed to maintaining the highest quality care; to advancing best practices in the delivery of palliative, end-of-life, and bereavement services and programs; and to increasing community awareness by engaging and cultivating supporters.
- *Financial security.* We will ensure the organization's long-term financial stability and security through integrity and responsible stewardship.
- *Leadership.* We will recruit and retain a committed, forward-thinking board, professional staff, and volunteers to fulfill the organization's mission and meet the needs of those we serve.

As we look toward the future, we can anticipate the delivery of end-of-life care to evolve. We are witnessing a confluence of trends and events such as a rapidly escalating aging population, demands of the boomers, lack of caregivers, legislation, evolving socio/demographic trends, hospice industry changes, and a challenging economy. In 2009, our board of directors commissioned an in-depth strategic planning process. Inclusivity was key to HOV garnering input from both our internal and external stakeholders through planned focus group meetings, visits to unique hospices, and discussions with the Institute of the Future. In June 2010, a Vision Conference was held, and later this year, HOV's board of directors will evaluate the information gathered and deliberate on the vision map that will take HOV into the next five years.

Please summarize the project/program for which these funds will be used

This is a request to assist HOV to provide free bereavement support to the children of hospice families and sliding-scale fee-based bereavement support to children in the community following the loss of a loved one. Receiving appropriate bereavement support after a loss helps grieving children through one of the toughest experiences any of us must face and has a positive, lasting impact on their lives. HOV's Community Grief Center offers the most comprehensive grief support program for children and teens in Santa Clara County. Services include anticipatory (or "pre-bereavement") counseling, individual and age-specific group counseling sessions, and special memorial and other grief-related events.

Current year budget: $10,756,200 **Projected budget next year:** Not yet available

Sources of your funding: 7 percent **Individual and Endowment;** 4 percent **Corporate, Foundation, and United Way;** 0 percent **Governmental;** 89 percent **Program Revenue** (Medicare, and private insurance reimbursement; copays and deductibles; and sliding-scale program fees for community bereavement clients)

Title of program for which contributed funds will be used
Children's Community Grief Center

Total program/project budget: $150,000 Is this a new project?
☒ Yes ☐ No

Please describe the project/program for which these funds will be used

The funding of this request will help HOV to provide free bereavement support to the children of hospice families and sliding-scale fee-based bereavement support to children in the community for 13 months following the loss of a loved one. Death and grief touch everybody at some point, but loss is particularly critical when it affects the lives of children and adolescents. Children grieve in their own ways, just like adults, and can have a very difficult time expressing their feelings in words. The effects of their grief, however, can be noticed in their school performance, their social interactions, and their level of self-esteem. Receiving appropriate bereavement support after a loss helps grieving children through one of the toughest experiences any of us must face and has a positive, lasting impact on their lives. HOV's Community Grief Center offers the most comprehensive grief support program for children and teens in Santa Clara County. Services include anticipatory (or "pre-bereavement") counseling, individual and age-specific group counseling sessions, and special memorial and other grief-related events, including the biannual Kid's Night Out Program hosted at Hospice of the Valley. Funds would be used for designated program expenses, with a particular emphasis on extending our grief support to greater numbers of low-income children and families in the community and purchasing art supplies for the age-specific support groups for children. Art and play therapy offer children an opportunity to express their feelings nonverbally in a safe and nurturing environment. Many of these children's beautiful and meaningful art projects are displayed in the Children's Art Therapy Room and give comfort to other children who are grieving. Based on this success, we are piloting an Art Show of pieces created by grieving clients that will premier on November 19, 2010. Sharing without words can help children to open up—both to themselves and to their families and friends.

HOV's Community Grief Center addresses the emotional, physical, behavioral, practical, and spiritual issues grieving individuals face. Activities include multiple components: one-on-one counseling, small peer support groups, phone consultations, volunteer support, educational literature (books, articles, newsletters), and special memorial and holiday events (the first holidays often being particularly painful without the loved one).

$150,000 in support in the 2011–2012 and 2012–2013 fiscal years would be used for the following expenses of the children's program in the Community Grief Center:

- Individual and group counseling sessions for 260 children and their families
- Kid's Night Out held biannually
- Training, curriculum development, and clinical oversight of interns and trainees
- Art therapy supplies, support group snacks, and refreshments
- Supplies and framing expenses for the Art Show (open to the community)
- Creation of a children's library
- Advanced professional development for staff clinical practitioners
- Technology (laptops, digital camera, photo-quality copier/printer)
- Volunteer training, support, and recognition
- Publications for parents of grieving children
- Newsletters produced quarterly
- Mailings and brochure production, printing, and postage
- Rent, utilities, and overhead

Why is this project critical in fulfilling your organization's mission and goals?

HOV's mission is to affirm dignity, hope, and comfort for those facing a life-limiting illness by providing compassionate palliative, hospice, and grief care. Grief support for both hospice families and the community at large is a vital part of our mission. Since grief can have a deleterious effect on family dynamics and stability, HOV is committed to helping entire families cope with grief and gain emotional resilience. Hospice of the Valley's Community Grief Center offers compassionate caring: condolence cards, one-on-one counseling, small peer support groups, phone consultations, volunteer support, educational literature (books, articles, newsletters), and special memorial and holiday events. With grief, one size does not fit all, and at HOV, we tailor each grief support plan to meet the family members' specific needs.

Our grief support services are designed to serve these personal and diverse needs. The variety of activities available helps to ensure that surviving loved ones, who come from many different cultural and spiritual backgrounds, have access to the grief support that will be most appropriate and beneficial. Each grieving individual is encouraged to choose the grief support program components that will be most useful and comfortable for them. A condolence card or a few phone calls may offer precisely the support one person needs, whereas ongoing attendance at a support group or visits to a grief counselor may help to transform another person's grief process and provide lifelong tools for coping with a multitude of feelings. Timely grief support not only helps the bereaved through the grief process but also leads to long-term emotional resilience and improved coping skills. This is especially true with grieving children, who often need nonverbal outlets to express their feelings of loss and sadness. Through individual counseling, age-specific children's support groups, and the art therapy program, HOV allows grieving children to acknowledge their emotions and begin the painful process of moving ahead without the lost loved one.

Who will benefit, and how many will be served?

HOV serves clients throughout Santa Clara County. Santa Clara County has the largest population of the Bay Area counties (1.7 million residents, according to the U.S. Census Bureau) and is one of the most diverse regions in the country: there is no majority population among the county's racially and ethnically varied inhabitants. HOV serves clients who are culturally and economically diverse and range in age from infancy to over 100 years.

Although Community Grief Center clients can be of any age, this proposed funding will specifically benefit children between the ages of 4 and 18 and their families/caregivers. The Community Grief Center is designed to support entire families coping with the loss of a loved one. Our research and experience working with bereaved children and youth have shown that grief affects children for the rest of their lives. In addition, there is an increased potential for long-term emotional trauma and negative consequences when a child loses a parent, and the majority of grieving children we serve at HOV fall into this category. When a child's parent has died, his or her center of security is shattered. This often leaves the child confused and angry. Any childhood loss, however, can have a lasting positive impact. Receiving adequate grief support at the time of a childhood loss has a positive and constructive impact on a child's ability to adjust to and minimize the effects of the loss. Comprehensive grief support allows for normalization of the child's grief process and development of new coping skills. By improving the child's self-esteem and helping to repair his or her confidence in the security of life, grief support provides a foundation for an emotionally healthy future.

Grief support for children has a long-term community impact. The consequences of neglected childhood grief often show up as attachment disorders, substance abuse, and even antisocial and criminal behavior. There is a clear and demonstrable effect on the community or society as a whole; the importance of timely grief interventions with grieving children cannot be overemphasized.

We project that HOV's Community Grief Center will serve 125 grieving children and teens in fiscal year 2011–2012, with a 10 percent increase in the 2012–2013 fiscal year (137 children and teens). This proposed funding therefore would subsidize grief support services to approximately 262 children and teens and their families.

What changes do you expect to occur as a result of the project?

As a result of their participation in grief support groups and individual grief counseling, children and teens served by HOV's Community Grief Center will be better able to cope with their grief and talk about it, they will function better in school and at home, and they will act out less and be able to give voice to more appropriate expressions of their feelings. Children who receive timely and appropriate grief support will be more resilient and better able to cope with loss and trauma later in life. Specifically, HOV will assist participants to

- Better understand their feelings and the grieving process
- Improve their emotional, spiritual, and practical coping skills
- Reduce isolation
- Improve their emotional and physical health, which decreases the likelihood of illness, drug or alcohol abuse, and truancy
- Improve family dynamics
- Accept the loss and reengage in life with a renewed sense of hope

Measurable objectives include

- Serving 262 children over two years through predeath sessions, one-on-one counseling, group counseling, and annual memorial and Kids Night Out events. The projected outcomes are as follows:
- At least 70 percent of group counseling survey results show improvement in social adjustment.
- At least 70 percent of group counseling survey results show improvement in school adjustment.
- At least 75 percent of clients at the end of individual counseling report that their goals have been met.

It is important to note that social and emotional adjustment typically continue to improve over a longer time period—usually beyond the time of our contact with the child or teen—as the coping tools they have been given here are used to help them through subsequent losses in their lives.

In what way is the project unique?

HOV's Community Grief Center is unique in that we have the most extensive art therapy program for children and teens and the only age-specific support groups run by licensed grief support professionals in Santa Clara County. Children who experience loss of a loved one require special care and attention. Our expressive arts program provides children and adolescents with vital activities and techniques that imbue loss with sense and meaning. Our approach is designed to instill in children and youth the necessary developmental assets and skills that increase emotional resilience and allow them to be successful at home and in school. A key element of our work with children is harm reduction. Children from higher-risk backgrounds often have experienced other developmental insults that may be exacerbated by the loss of a loved one. If a child receives the proper types of intervention after the loss of a loved one, however, there is a greater likelihood that no additional harm will occur to the child. This is an important prevention-based aspect of our work with grieving children and teens. The program is also unique in that the majority of children suffering from grief go unnoticed in many settings ranging from school to foster care. In addition, when one considers other systemic barriers, such as lack of access and varying cultural and socioeconomic situations, it is apparent that there is not enough of this vital service in our community. The inclusiveness and availability of HOV's comprehensive program make it unique in a community where there is a paucity of affordable and accessible grief support services.

How does it fill a need that is not being met in any other way?

HOV's grief support program for children and teens fills a need that is not being met by schools or other agencies in the Santa Clara Valley. Most schools are not equipped to work with grieving children, many families lack health insurance or have insurance that does not cover sufficient grief counseling for children, and most children's grief support groups offered by other organizations in the area are run by volunteers instead of licensed therapists. HOV's Community Grief Center provides professional, culturally sensitive, low- or no-cost grief support to a large community whose bereavement needs are growing rapidly. We would like to continue expanding our Community Grief Center to help meet the growing need.

Who are the organization's current/proposed collaborative partners in implementing project?

We collaborate with a number of other children's bereavement programs in the Bay Area, such as the Santa Cruz Hospice in Scotts Valley, which serves Santa Cruz County, and KARA in Palo Alto. Although most of the children in our program come from the families of our hospice patients, we also serve community families on a sliding-scale basis. These families can be referred to us by family counselors, churches, schools, etc.

Medicare requires only that hospice providers send condolence cards to families when their loved one dies. HOV provides far more comprehensive grief support services, and we rely for program design and expansion not only on the expertise of our own clinical professionals but also on the experiences and best practices of a wide variety of bereavement service providers. Our Community Grief Center program staff belong to the Northern California Bereavement Coordinators, the California Association of Marriage and Family Therapists, the American Art Therapy Association, the National Association of Social Workers, and the National Hospice and Palliative Care Organization.

We have collaborated with KARA over the past three years to participate in its annual grief conference. Additionally, we are partnering with Los Madres, a support network for women with young children, to offer a yearly workshop at the HOV campus. The inaugural event, entitled, "The Elephant in the Room: Children, Grief, and Loss" will be held on October 28, 2010. Our goal is to reach out to women in support of their children during difficult life transitions, that is, divorce, death of a loved one, or relocation.

We would like to expand our collaborative grief education and outreach to help serve children in the community who have virtually no one to advocate for their grief support needs. HOV has identified an ongoing need to reach grieving children and teens through their schools. Although we make ourselves readily available to schools and have done crisis work on some campuses (most recently at XYZ School), it has been challenging to provide grief support to certain children and teens without greater access to their schools. Children who have lost parents are sometimes living in foster care or group homes and may not have transportation to support groups here at HOV. Having access to these children at their own schools would facilitate increased outreach and service delivery to this population. Most schools have their own on-campus or district counselors, and we think that it would be very beneficial for us to offer some educational workshops on grieving children to improve the available campus-based grief support.

Many of the school counselors do not have the time, skill, or capacity to focus on the needs of grieving children. We have bereavement hours open for schools with a grief counselor who specializes in children's grief, and our goal is to leverage the hours to provide education and support on school campuses, in shelters, and in group homes.

How will you evaluate the results of the project?

Outcomes will be measured through parents' self-report data in evaluations of the grief support services. Parents or caregivers will be asked to complete this evaluation after each six-week group series and again 12 to 13 months after the death of their loved one. Surveys of group participants consistently demonstrate that we are achieving our program goals. Our bereavement clients tell us repeatedly, "I couldn't have done it without you."

Additional evaluation is more subjective, with visiting social workers assessing the children's and teens' increased understanding, coping skills, and need for follow-up support.

If this is an ongoing project, how do you plan to sustain it in the future?

Ongoing fundraising will always be necessary to support HOV's high-quality Community Grief Center and its activities for children and teens. Although Medicare mandates that all hospices provide bereavement support, there is no Medicare reimbursement for bereavement services. Program fees from community bereavement clients cover only 8 percent of the total program costs. Support will enable us to support our clients more effectively as they move through their individual grief processes and will help HOV to leverage additional funds to support the program as it grows.

Please attach the following:

- Current year budget, including revenue and expense
- List of your primary contributors
- List of grants or donations pending or secured for this project
- List of your board of directors, including affiliation and contact information

Sample Long Grant Proposal

Children's Discovery Museum of San Jose

Grant Proposal to the XYZ Foundation for Biomedical Sciences for BioSITE (Students Investigating Their Environment)

Children's Discovery Museum of San Jose (CDM) appreciates this opportunity to request renewal funding for *BioSITE* (Students Investigating Their Environment) in the amount of $25,000 for the upcoming 2007–2008 school year. One of the museum's most recognized and innovative education initiatives, *BioSITE* enhances the science education of elementary and high school students and their teachers through hands-on inquiry activities and biomonitoring of the Guadalupe River watershed. As we prepare for our fifteenth year of program implementation, CDM is proud of the interest in *BioSITE* by the regional and national education communities. Teachers throughout the country, even internationally, regularly access *BioSITE* online through the CDM Web site, and there is a growing interest in replicating this model environmental education and stewardship program in communities across the nation. As the following pages will show, *BioSITE is changing the way students learn and teachers teach science.*

Children's Discovery Museum of San Jose

Children's Discovery Museum of San Jose is an arts, sciences, and humanities museum for children and youth, their families, caregivers, and educators. For nearly 17 years, CDM has been providing "hands-on/minds-on" opportunities for child-centered investigation and interpretation of the local natural, cultural, and built environment. In addition to 150 permanent exhibits, a traveling exhibition hall, two art studios, a theater, and an early learning center, the museum leverages the surrounding natural environment through a 1,200-square-foot outdoor *Kids' Garden* and a cadre of in-depth science programming focused on the adjacent Guadalupe River.

Since opening our 52,000-square-foot facility in 1990, nearly 5 million visitors from diverse local communities have been served by our mission "to unleash a sense of wonder, creativity, and joy in learning through playful exploration and inquiry." They have participated in rich process-oriented, inquiry-based arts and sciences programming, have joined together in celebration during museum cultural festivals and performances, and have benefited from the museum's strong agenda of service to the community. Our location in downtown San Jose, bordered by low-income, primarily minority neighborhoods, ensures that CDM reaches those least likely to have access to enriched educational opportunities. Of the more than 300,000 young children and adults who visit the museum each year—from Santa Clara, Alameda, Santa Cruz, San Benito, and San Mateo counties—85 percent come from areas within the immediate 30-mile radius. Twenty-one percent of our audience is Latino, 17 percent are Asian/Pacific Islander, 4 percent are African American, and 58 percent are Caucasian. Through funds raised for our *Open Door Policy*, 23 percent of our visitors receive free or subsidized admission. Beyond CDM's walls, approximately 50,000 take part in free museum programming through in-school programs, community festivals, and neighborhood events.

With a particular commitment to supporting the youth and teens of San Jose, the museum offers multiple service-learning, mentoring, and volunteer opportunities for middle and high school students. Through programs such as *Summer of Service, Discovery Youth*, and *BioSITE*, the museum supports school academic achievement while working to empower participants with the tools to positively affect their lives, communities, and local natural environment. In 2001, programs such as these attracted the attention of First Lady Laura Bush, who awarded CDM the Institute of Museum and Library Services (IMLS) National Award for Museum and Library Service.

BioSITE

In 1993, Children's Discovery Museum of San Jose launched *BioSITE*, a comprehensive environmental education program then involving fourth and fifth grade students from two downtown San Jose elementary schools in what was envisioned as an ongoing long-term field study of the Guadalupe River watershed at a site adjacent to the museum facility. Completing its fourteenth year, *BioSITE* now serves nearly 1,000 students annually, having expanded to include significant program components for high school students, college and pre-service teachers, an after-school program for elementary school students called *ACE* (After Class Experience), teacher training, and expanded program dissemination through the *BioSITE* Web site.

BioSITE elementary school students engage in ongoing field studies along the banks of the Guadalupe River (an urban stream running adjacent to participants' schools and through the communities in which they live) to learn about the watershed, they conduct water quality tests, develop basic field study techniques, and pursue inquiry-based science investigations. The program's curriculum includes biweekly field work and water quality monitoring (testing dissolved oxygen, turbidity, pH level, dissolved solids, temperature, and rate of flow) conducted by fourth graders working in groups of four to six students with trained high school students and preservice teachers. *Approximately 50 percent of the elementary school students enrolled in BioSITE complete their field work as part of the academic school day, whereas the other half of participating students engage in BioSITE after school hours through ACE.*

Now in its fourth full year of operation, *ACE is a proven and successful component of BioSITE*. Nearly 1,000 elementary school students now participate in the program annually, participating in three unique field sessions along the Guadalupe River. *ACE* high school–age mentors are recruited from throughout the region through the Youth Group, offering a diverse population the opportunity to earn service-learning credit while gaining science and leadership skills. They receive weekly training on *BioSITE* content and gain skills in mentoring younger students while strengthening their own leadership abilities, confidence, and understanding of environmental science.

BioSITE high school students take part in a year-long *BioSITE* science elective course and receive teacher training tips, field science instruction from professional scientists, and practical work in the field with elementary school students. Two high schools—ABC High School and Red High School—now participate in the *BioSITE* high school program. It should be noted that ABC targets low-achieving urban minority students, explicitly preparing them for college success. In fact, a significant number of ABC graduates will be the first in their families to attend college. Many of ABC's students now mentor elementary school students from the very same schools they attended as kids, making them not only role models for academic excellence but tangible examples of what is possible for this next generation of young children.

Thorough preparation prior to a *BioSITE* experience is critical to having a successful outcome for the students and classroom teachers. To that end, CDM has developed a suite of resources in supporting its *facilitators and classroom teacher participants*. Program facilitators and teacher participants receive professional development through a three-day *BioSITE* Teacher Institute at the start of the school year. *BioSITE* structures its professional development around science content as well as science pedagogy. Further training for facilitators comes in the form of twice-monthly training sessions in preparation for upcoming field days. For classroom teachers with access to the Internet, the CDM Web site includes program descriptions, access to useful biomonitoring data with a unique interactive design, and curricular materials in a downloadable format, saving classroom teachers valuable time in preparing for their visit. *BioSITE* facilitators also provide ongoing support for classroom teachers in the form of additional teaching days at the school site designed to augment a teacher's classroom-based goals. Additionally, CDM offers significant stipends to teachers and facilitators to attend the *BioSITE* Teacher Institute.

Over the course of its 13-year history, *BioSITE* has enjoyed numerous successes and continued growth. *Demand for the program continues both locally and from schools and institutions outside the state of California*. At a recent conference of the Association of Science and Technology Centers

(ASTC), a *BioSITE* training session was oversubscribed. Participants representing museums, schools, and other NGOs throughout the United States requested information about the ways they might adapt the *BioSITE* model for use within their own communities. The positive feedback from this event and at other similar conferences signals a need for broad dissemination and replication of this unique learning system.

Specific Request

CDM respectfully requests a grant in the amount of $25,000 from the XYZ Foundation for Biomedical Sciences to support ongoing program delivery and dissemination of *BioSITE* in the coming year. Specifically, the support of the foundation will further the achievement of the following measurable objectives:

Improving school science learning for students:

- 400 elementary school students will participate in eight in-depth field study sessions at the museum's and Red High School's Guadalupe River research sites during in-school hours.
- Up to 1,000 elementary school students will attend three ACE after-school field study sessions at CDM's Guadalupe River research site (2½ hours per session), conducting field work and data collection and engaging in inquiry-based learning.
- 50 students from Red and ABC High Schools will complete a year-long *BioSITE* science elective course administered in their schools, take part in field work exercises and hands-on investigations with practicing scientists, and serve as mentors to elementary school students during watershed field research.
- 12 Youth Group members—high school and college students and pre-service teachers—will serve as facilitators for the ACE program, each receiving 100 hours of in-depth content training (including work with practicing scientists) and field work with elementary school students.
- CDM will complete a draft of specialized *BioSITE* curriculum for high school students, inclusive of mentoring techniques for each field experience.

Supporting teachers through professional development:

- 30 teachers will participate in a three-day *BioSITE* Teacher Institute focused on *BioSITE* implementation training, including science inquiry processes and assessment techniques, field research experiences, and an exploration of program-related scientific content. Ongoing support will include progress meetings with CDM *BioSITE* staff, monthly updates, and participation in professional development seminars.
- CDM will develop content and an implementation plan (to include teacher stipends and transportation assistance) for a three-day Teacher Institute for educators throughout the country wishing to carry out a *BioSITE*-modeled program in their schools and communities.

Project dissemination:

- CDM will continue to support students, teachers, scientists, and the interested public through *BioSITE* Online (www.cdm.org/BioSITE), by providing full access to the *BioSITE* curriculum, field research data, and resource links and a discussion forum for *BioSITE* educators. We will create additional online support materials for teachers in other geographic locales wishing to use and modify the *BioSITE* curriculum.
- CDM will complete translation of essential elements of the *BioSITE* curriculum into Spanish and Vietnamese, making it fully available on *BioSITE* Online.
- CDM will showcase *BioSITE* at special museum *BioSITE* days and events serving approximately 4,000 children and adults annually and through presentations at regional and national conferences such as Youth Quest (Headlands Institute) and the National Service Learning Conference.

Evaluation

CDM recognizes the value of evaluation and has made concerted long-term efforts to examine the impact of its programs through the development of assessment tools, both in-house and with the aid of outside evaluators, including

- Before and after interviews, surveys, projects, and academic style tests are administered at all levels of participation from students to teachers and visiting scientists. These tools are used to establish changes in attitude, awareness, behavior, and content understanding (specific examples of tools available on request).

- Program observations
- Journal assessment: Student participants enter their research data, including water test results, observations of river environment, and wildlife sightings, in their journals. The journals also serve as an essential tool students use to create a record of their inquiry processes (lists of questions, revelations, descriptions of experiments and results, drawings, etc.).
- Written and lab-style tests are used at the high school level.
- High school attendance and enrollment records for the *BioSITE* program are examined from year to year.

Evaluation of *BioSITE* enables CDM to set priorities and goals for the coming years and build on existing strengths to address any shortcomings of the program.

Conclusion

The science education field, including the National Science Education Standards, emphasizes the important role of inquiry and student-centered learning, yet few in the present generation of teachers have experienced these techniques in their academic training. Responding to both well-documented and emergent science education needs, CDM is helping, through *BioSITE* to reform science education for students and teachers. The *BioSITE* approach to teaching science makes direct the link between the abstractions of the classroom and the real phenomena of the natural world. *BioSITE* not only promotes understanding but also fosters a sense of individual responsibility for conservation and wise resource use. Results from research indicate that youth clearly benefit from hands-on, inquiry-based environmental education (Barnett et al., 2006; Fusco, 2001; AIR, 2005). When hands-on environmental education is integrated into the school curriculum, student behavior improves, school attendance improves, and academic performance rises (SEER, 2001; AIR, 2005). We hope the XYZ Foundation will join us again this year in supporting this model science education program for students and teachers.

Proposed Budget	
Expenses	
Personnel:	
Director of education (35% of time)	$23,500
BioSITE manager (100% of time)	52,000
ACE coordinator (100% of time)	24,500
Facilitators	13,500
Taxes and benefits @ 22.5%	25,538
Total personnel	$139,038
Other Expenses:	
Teacher stipends	$ 13,500
Scientist stipends	2,400
School bus transportation	17,000
Evaluator/translation consultation	12,000
Equipment and supplies	17,000
Printing (field journals)	5,000
Conferences and meetings	2,250
Student participation in national conference	2,000
	$71,150
Grand Total	$210,188
Income	
Current sources of support for 2006–2007:	
Individual donor	$125,000
XYZ Foundation	25,000
A Corporation	20,000
City of XXX Watershed Program	5,000
	$175,000
Committed sources for 2007–2008:	
Individual donor (annual donor)	$160,000

A corporation (annual donor)	20,000
	$180,000
Pending Sources for 2007–2008:	
City of XXX Watershed Program	$ 5,000
XYZ Foundation	25,000
	$ 30,000
Total Committed or Pending 2007–2008:	$210,000

Key Contacts

Name
Associate Executive Director
Children's Discovery Museum of San Jose
Address
Telephone and extension
E-mail address

Name
Director of Education and Programs
Children's Discovery Museum of San Jose
Address
Telephone and extension
E-mail address

Name
Environmental Science Educator
Children's Discovery Museum of San Jose
Address
Telephone and extension
E-mail address

Sample Long Grant Proposal

Bay Area Glass Institute

Proposal to Arts Foundation

1. Describe your organization, when it was founded, and its mission, programs, and activities.

The Bay Area Glass Institute's (BAGI) mission is to make the celebration of glass art accessible to all and to provide continued artistic and educational growth to artists, patrons, and the community.

BAGI was established in 1996 and is the only nonprofit, publicly accessible glass studio in the Santa Clara County. BAGI's facility is housed in a 4,000-square-foot space located in San Jose's Japantown district. It is an ideal location for artists to develop, exhibit, and sell their artwork. Having a fully functioning studio allows BAGI to bring artists, collectors, and the public together at the moment of artistic creation. The studio also enables BAGI to teach hands-on classes for beginning through advanced students in multiple glass disciplines.

BAGI's three core programs align with its mission to make glass art accessible to all:

1. *Glass Art Education Program.* FY08 class enrollment was 673 seats (up 40 percent from FY07). The increase is linked to the purchase of new equipment/tools. The new purchases allowed BAGI to offer a broader spectrum of classes. BAGI will grow its FY09 class enrollment by 10 percent from FY08. The increase will be accomplished by
   a. Augmenting core curriculum with specialty and master-level classes
   b. Increasing BAGI's online presence on social networking sites and community calendars
2. *Glass Art Facility Enhancement Program.* BAGI asked students and artists, "What would prompt you to use BAGI's facility more?" The top five responses centered around studio enhancements. Specifically,
   a. Better tools
   b. Bigger glass garage

c. A second powder booth
d. A second fusing kiln
e. Better cold-working equipment

BAGI listened to its students and artists. It chose "Studio Enhancement: New Tools and Equipment" as its annual auction's fund-an-item (FAI). Through auction patrons' FAI donations, BAGI was able to update existing equipment and purchase new equipment and tools. Studio usage increased 5 percent in FY08 because artists and students were able to expand their craft with the new equipment/tools. BAGI's 2009 enhancement goals include reconfiguring parts of the studio to better use its workspace.

3. *Glass Art Public Access Program*. BAGI's annual public access programs include the Great Glass Auction, Great Glass Pumpkin Patch, Third Thursday Series, Visiting Artist Series, and Winter Wonderland. Solid planning, budgeting, and marketing, along with strong artist support, allowed BAGI to create well-attended, publicly accessible events. In FY08, over 30,000 attendees enjoyed free demonstrations, lectures, and exhibits put on by BAGI. BAGI also participates in "one-time-only" public access events/programs. For example, in FY08, the City of San Jose invited BAGI to exhibit at the City's Windows Gallery located in San Jose's City Hall. BAGI was honored to do so with its "Out of the Fire" exhibit that featured 30+ artists' artwork. BAGI intends to grow its public access programs by 10 percent in FY09. The increase will be accomplished by
   a. Fostering and improving existing programs while exploring new programs to reach a broader Silicon Valley–based audience
   b. Increasing BAGI's online presence on social networking sites and community calendars

2. Describe how your programs are received by participants and audiences.

In general, BAGI programs are well received by participating artists and program attendees. BAGI is able to gauge its program(s) successes by

1. Reviewing the responses/comments to its program(s) surveys
2. Increasing its program(s) attendance
3. Increasing its program(s) revenue

For example, BAGI saw a 40 percent increase in its class enrollment from FY07 to FY08—that's an increase from 483 students to 673 students. The increase was due in part to facility improvements and the addition of more classes. The Great Glass Pumpkin Patch Exhibit and Sale 2008 held at the Palo Alto Art Center is another example of a well-received BAGI program. The event

1. Featured over 40 artists (up 20 percent from FY07)
2. Saw 5,000+ attendees (up about 8 to 10 percent from FY07)
3. Sold 5,100+ pumpkins (up 10 percent from FY07)
4. Launched an online survey that garnered 900 responses in less than three months.

3. Describe the involvement of artistic personnel in program planning and implementation.

BAGI is continually asking for feedback from our instructors, artists, students, event patrons, and volunteers. All students complete evaluation forms after each class that are reviewed by staff and instructors to improve our offerings. BAGI holds postmortem debriefs and artist meetings after every one of our events to make them better. Results from this feedback are reviewed regularly by the board at the monthly meetings.

4. List your organization's key artistic personnel.

Name	Role	Background
	Executive director (full-time employee), instructor, volunteer	BA in Sociology, X University; MA in Parks and Recreation, X University; former Sales and Vendor Relations Director; Glass Art Society member; glass and mixed-media artist
	Operations manager (full-time employee), volunteer	BS in Advertising, X University; Glass Art Society member; former Director of Channel Marketing; currently enrolled in various art and business management classes; glass and mixed-media artist

Studio manager (part-time employee), instructor, volunteer	Former IT Project Manager; Glass Art Society member; currently enrolled in various art and culture classes; glass and mixed-media artist
Torchworking teacher (independent contractor), volunteer	Member of the International Society of Glass Beadmakers and the ABC Bead Society; former program manager; glass artist
Fusing teacher (independent contractor), volunteer	Art instructor for local art organizations; founder and president of the XYZ Artist Alliance; glass and mixed-media artist
Glassblowing teacher (independent contractor) volunteer	BA in Art History and English, X University; Member of ABC Museum X University; Member of ABC Museum of Glass "Celebrity Solstice" glass-blowing team; glass and mixed media artist
Glassblowing teacher (independent contractor) volunteer	BFA, X University; Instructor at ABC Museum of Glass; glass and mixed-media artist

5. Describe any particular audience you serve and the benefits your organization provides to Santa Clara County.

BAGI is a focal point for local glass artists, collectors, students, and the public to the 3,000-year-old craft of working in glass. Its publicly accessible, San Jose–based studio allows artists and students to develop their glass art skills locally. All but one of its annual events are free to attend and all are welcomed. Collectors and customers from all over Santa Clara County attend BAGI events and shop in its gift shop and gallery.

BAGI's public demonstrations, exhibits, and lectures teach the community about the history, science, and techniques of glass art. Past Santa Clara County–based organizations that have attended demonstrations, exhibits, and lectures at BAGI include the Boy and Girl Scouts, Arts Express students from the ABC School District, various home school associations, and various senior citizens' groups (e.g., the downtown seniors groups, Palo Alto seniors groups, and Saratoga seniors groups).

BAGI hosts regional, national, and international guest glass artists to teach and conduct demonstrations and lectures at its studio. There isn't another recognized 501(c)(3) nonprofit glass art facility that provides this service/opportunity to Santa Clara County residents.

BAGI partners with fellow recognized 501(c)(3) nonprofit organizations in Santa Clara County to produce publicly accessible glass events, including:

1. Great Glass Pumpkin Patch, now in its thirteenth year, cohosted with the XYZ Art Foundation
2. Glass Farmers Market with the XYZ Artists Alliance
3. Spirit of Glass 2009—BAGI in talks with XYZ Business Association to define BAGI's participation

The Great Glass Auction, held in San Jose, showcases some of the finest glass art produced by our local artists, as well as bringing art into Santa Clara County by some of the best glass artists in the world.

6. Describe your long-range planning process and annual goals. What are your major challenges, and what are your strategies to address them?

Over the past two years, BAGI has been focused on becoming financially stable. In 2006, through operating cost reduction and volunteer support by the board of directors, BAGI was able to end the year generating a small surplus. An executive director was hired in February 2007, and he has improved our facilities and our financial position. BAGI has been able to retire the equipment lease on our furnaces, repay 25 percent of the loan made to BAGI in 2005, and again end the year with a small operating surplus. In 2010, all existing debt of the organization will be paid. This fiscal discipline has been created by a tight working relationship between the board, the treasurer, and the executive director during the budget creation process and the ongoing maintenance of our program and spending plan.

Creating an updated long-term strategic plan that will look over the next 10 years is vital to the organization's ongoing impact. The board has held multiple discussions to decide on the critical few priorities to address in that long-term plan. Those key priorities are

- Funding and finding a permanent, owned facility
- Creating a model for a more environmentally friendly glass studio (tied to the first item)
- Funding and hiring a dedicated fundraising position within the organization
- Expanding our reach into the Bay Area community for participation in classes and programs

The first and second priorities are being accelerated due to changes in the external environment. Those changes—rising energy costs, the awareness of our studio's environmental impact, and the belief that it is possible to improve that impact—have created an opportunity for as well as a threat to BAGI's success. This issue is not unique to BAGI because all "fire arts" facilities are facing the same dilemma in the United States. As energy costs increase, the increased fees for classes and studio rental time are making it prohibitive for beginning and emerging artists to afford to practice their craft. The staff, with help from passionate board members, is currently researching potential solutions, and those will be used as input into BAGI's strategic plan.

Those four important elements of the strategic plan need to be fleshed out with significant work and input by the board, staff, external experts, and the community. In the fall of 2008, BAGI secured a technical assistance grant from the City of ABC to develop our strategic plan. BAGI has hired a consulting firm to assist in the project. It is anticipated that the plan will be completed in May 2009.

7. Describe the roles of volunteers in your organization.

BAGI has always depended on our volunteer base to support the activities of the organization. At our major events, the volunteers help with docent presentations, manning sales and packing stations, and assisting our patrons and the public. The annual BAGI Auction Committee consists of all volunteers, and they create the marketing programs, recruit artists for donations, invite and follow up with the patrons, and assist with staging of the event. Countless hours of some 40 volunteers are involved. BAGI artists and students volunteer their time to assist visiting artists with presentations, assist artists in the creation of their work, and provide support activities that make the events successful. Our artists and students also volunteer their time to assist with the upkeep and repair of the BAGI facility. The Great Glass Pumpkin Patch involves over 80 volunteers to set up, conduct the sale, and take down the event.

8. How do you publicize your programs to the community?

BAGI publicizes its programs via the following vehicles:

Print. Public relations outreach leading to articles in the *ABC Newspaper* and multiple other local papers, ads in local publications for select events, mailing of post cards for events to our 5,000-plus database, and the distribution of quarterly newsletters, class schedules, and flyers at events

Electronic. Maintaining an up-to-date Web site with information on all BAGI activities (in 2008, www.bagi.org had over 800,000 visits); sending event e-mails to our patron database; posting to various Web sites, including Artsopolis, Metro Online, and San Jose Mercury News online calendar; search engine marketing ad campaigns on Google; creation of Facebook pages for key events. In 2008, BAGI created the www.GreatGlassPumpkinPatch.com Web site to specifically promote the event and participating artists. During the three months preceding the event, the site received over 400,000 visits.

Radio. Major events are covered by public service announcements (PSAs), with over 14 stations in the Bay Area providing air time.

TV. Major event coverage by KNTV Cable, 15 PSAs, and KICU-TV36 Sunday morning segments.

9. Describe any plans for expanding your audience and/or participant base.

BAGI's 2009 program plan was crafted after the staff and board members reviewed its 2008 programs, financials, challenges, successes, and feedback from students, artists, and patrons. The reviewed information assisted the board in creating a plan that will foster BAGI's 2009 core programs. BAGI will increase resources on marketing programs to attract more students, artists, and patrons to its core programs. The increase will be dedicated to creating outreach collateral that is easily distributed to BAGI's patron base of 6,000 people, local schools, libraries, community centers, and continuing education programs. The collateral will be designed to assist BAGI in meeting its program goals by

- *Featuring information about its classes.* BAGI taught 670 students in 2008 and plans to maintain current class levels in 2009 despite the current economic conditions.
- *Describing the benefits of renting time in BAGI's studio.* BAGI's 2008 facility was used by over 40 artists. They rented 2,400 hours of time in the studio. It is BAGI's intent to increase facility usage by 5 percent and aid artists in meeting their artistic goals.
- *Offering highlights of BAGI's free, publicly accessible events.* In 2008, 22,400 people attended free events. It is BAGI's intent to increase event attendance by 10 percent. The collateral will include highlights of BAGI's 2009 events calendar.

This new collateral and outreach is focused on increasing BAGI's patron database by over 10 percent in 2009.

10. How are strategic planning and decision-making carried out in your organization? What is the role of the board, staff, volunteers, and advisory groups in this process?

The BAGI board meets monthly to review the organization's financials, programs, and operations to ensure that we are on track and provide input to the executive director. The board also discusses BAGI's execution on critical projects and programs with the executive director.

In late 2008, BAGI received a grant from the City of ABC to develop a strategic plan. With the assistance of ABC Consulting, input is being solicited from the board, staff, artists, volunteers, patrons, members of the public who have attended BAGI events, and similar art organizations. The data will be reviewed by the board, and the plan will be developed. Completion date for the strategic plan is May 2009.

11. Do you have a budget deficit or surplus? What are your plans for dealing with it?

In 2008, BAGI incurred a small operating deficit ($9,331) due in part to the reduction in anticipated revenues from the Great Glass Pumpkin Patch in October 2008. This was covered by reserves from previous years. Staff and the board are reviewing revenues and expenses monthly to adjust to the current economic conditions.

Sample Long Grant Proposal

Homeless Agency for Intact Families (HAIF)

Capital Campaign Grant Proposal
Application to XYZ Corporation

XYZ Corp Funding Focus: Children, families and communities, Internet safety and security, green building
Funding request amount: $15,000 grant

Funding for: "Tech Savvy Training" in computer center
Contact name:
Contact telephone:
Contact e-mail:
Contact fax:

Organizational Information

Founding date: 1986

Mission statement: To provide temporary shelter and supportive services for homeless families in ABC County to empower them to move out of homelessness and into self-sufficiency in our community.

Population served: Homeless, intact families in ABC County

Staffing levels: Full-time employees (FTEs): 15; part-time employees (PTEs): 7; volunteers: hundreds

Operating budget last FY: $1.9M

Top two signature programs: Family Shelter and Group Home

Past successes and honors: Recognized by the City of EFG for our transitional housing and employment assistance programs. Recognized by the Association of Fundraising Professionals (AFP) for our volunteer board and fundraising leadership.

Geographic Scope

Regional: Within 10 miles of XYZ Corporation

Partner Involvement

HAIF is working with the City of EFG and in partnership with the local commission.

HAIF is proud to be involved in the work of the local commission dedicated to ending homelessness and solving the affordable housing crisis.

The work of the commission is both grounded in research and seeded with optimism in its ambitious goal of ending chronic homelessness in ABC County by the year 2015. The commission is a strong ally in our effort to build a new HAIF housing residence. As the availability of affordable permanent housing overtakes the need for temporary housing of homeless families, the flexible design of the new HAIF residence will allow us to shift to meeting the need for single-occupancy units of permanent housing with onsite wraparound services.

The XXX Foundation has been partnering with us for many years, providing us with "food and shelter" funding plus three organizational effectiveness grants. We are much appreciative of these generous gifts to date, and we see our partnership continuing well into the future. We are awaiting word on a significant grant request made to XXX.

We are actively submitting applications to preresearched foundations according to their guidelines and submission schedule. We hope to receive funding from the YYY Foundation for our "green" initiatives, and we have been given encouragement by ZZZ Foundation and WXY Foundation.

Numerous local philanthropists have already made generous donations to the campaign, including the XYZ Charitable Trust. We have 100 percent participation by the HAIF board, past and present, the Capital Campaign Steering Committee, the HAIF Advisory Council, and the staff.

Project Specifics: Tech Savvy Program

Opportunity, Transformation, and Impact

HAIF has been transforming lives since 1988, operating a shelter and support programs for homeless families in a converted warehouse in the City of EFG. HAIF is the only local temporary shelter that serves intact homeless families. While Homeless Agency for Intact Families provides exceptional services for this vulnerable population, the City of EFG mandated 10 years ago that the shelter move because room sizes are below code.

Since we must move, Homeless Agency for Intact Families has an opportunity to build a new residence for homeless families as part of a new transit village designed around a future mass transit station. HAIF is launching a multi-million-dollar effort to construct a state-of-the-art green short-term housing complex exclusively for families with children.

Once completed, the building will be one of a kind, combining our short-term housing program with a larger affordable housing development. It will create a seamless program aimed at moving families from temporary housing into permanent, affordable rental housing while encouraging long-term self-sufficiency.

Creating Lasting, Visible Change

The emphasis at Homeless Agency for Intact Families is and will always be on creating lasting change. We will continue to offer training and enrichment programs and workshops in addition to our well-respected Transitional Housing Program. Hundreds of volunteers from faith-based groups, service groups, and corporate partners prepare and serve meals, take children on outings, provide clothing and supplies, tutor, teach skills, etc. throughout the year. Without the volunteers, the work of HAIF would be greatly diminished.

To complement XYZ Corporation's goals to educate the community about safe Internet usage practices and protecting valuable information online, we propose joining forces so that XYZ Corporation's trained staff can teach a *Tech Savvy Training Program* to the children and adults of the current HAIF Shelter and the new HAIF Family Residence. The program, instructors, and materials would be furnished by the XYZ Corporation. *The $15,000 grant would go to fund the area in the new residence where the program would be taught.*

As the XYZ Corporation knows, one in three young people are enticed online in one form or another. Safe and secure Internet practices need to be learned and continually updated to prevent horrible incidents or system invasions that can destroy, alter, or make use of confidential information. The XYZ Corporation partnering with HAIF would enable us together to reach a sector of our community most vulnerable to fraud and enticement.

HAIF's commitment to build a "green" facility is not one to be taken lightly. This commitment added significantly to the overall budget. But the executive director of HAIF and the board of directors agreed that it was the right thing to do "for the greater good" of our community and the universe.

Measurement and Evaluation

Goal of the *Tech Savvy Training Project*: Understand and encourage behavior changes of residents of HAIF facilities toward Internet safety and security.

Tech Savvy Objectives:

- Create awareness of the issues around lack of understanding of Internet safety and security.
- Create interest in the benefits of changing certain online and computer practices.
- Create a desire to make simple changes that can have long-range beneficial effects for children, parents, and families.
- Create action plans for each family that can serve as a positive incentive when they leave the Shelter or new Residence and move into a positive part of the community.

Methods, Activities, Strategies:

- Weekly class for adults
- Weekly class for junior high and high school students
- Weekly class for grade school students
- "Tips" sheets and coloring books to be available in hard and soft copy
- Classes to be taught by XYZ Corporation's trained employees
- Tie the learning modalities to Project Cornerstone asset measurements through before and after measurement tools.
- Computer center in new Residence to be outfitted with Internet security software (could be contributed by XYZ Corporation).

Success Looks Like:

- Adults will be able to demonstrate and articulate key learnings from the trainings.
- Children and youth will be able to demonstrate and articulate key learnings from the trainings.
- Program staff will be able to measure asset growth through before and after measurement criteria.
- Staff and management will be able to report to funding sources the progress and successes of the families coming through the HAIF Shelter and the new HAIF Family Residence.
- A cadre of XYZ Corporation volunteers also will volunteer for other aspects needing assistance at HAIF.
- XYZ Corporation becomes one of our key partners and friends through the building of the new Residence and beyond.

Evaluation Information:

- Will be communicated to funding sources, influentials, donors, and volunteers via appropriate communication devices such as
- Annual report
- Reports to the city and state
- Newsletters
- Posters created by the children for visitors to see
- Public relations efforts for sponsors, donors, etc.

Sponsor Acknowledgment and Recognition Plans:

- Signage
- Public relations efforts
- Event mentions
- Newsletter mentions
- Class materials
- Others that would be mutually beneficial

Project Budget

The total cost of the new HAIF Family Residence, land and all, is $16 million dollars. Of this total, $11 million has been committed or identified from public monies and other sources. We are currently in the middle stages of our $5 million dollar private monies campaign. We have raised nearly half the funds.

Measuring Success

Our commitment to creating lasting change for vulnerable families leads us to provide quality services after families leave temporary housing. It is our practice to track the status of our families through our care process—intake, care plan, and after-care case management—and through our successful temporary and permanent housing referral program. Some families return to volunteer for others and give testimony to their success.

Dimensions such as the ability to get a job, go back to school, and rent an apartment or home, as well as physical and emotional factors, all become a part of the overall program to track progress.

We have tracked a 70 percent success rate overall of families leaving the Shelter and going into permanent housing. If those families go through our Transitional Housing Program, the percentage jumps to 87 percent. These are admirable statistics compared with other programs.

We estimate that 60 percent of the families with us for 90 days have one parent working, and over half have one parent in some type of schooling. All the children age three and over staying at the HAIF Family Shelter are in day care, preschool, or school.

Organizational and Financial Capacity

To augment our extensive experience serving homeless families, we have teamed up with XXX to leverage its experience in affordable housing development. XXX is acting as our project manager.

We are rapidly acquiring funds for this project. The total cost for this project is estimated at $16 million based on a detailed development budget.

We already have most of the $11 million of public financing committed. Our efforts to raise the remaining $5 million are going well in this early major donor phase of the capital campaign. Please refer to the "Public and Private Financing Sources" spreadsheets that are included.

The capital campaign plan is also included for your information. The plan is built on the basis of a third-party planning study, with recommendations, goal, objectives, strategies, timeline, and budget included. We have retained outside counsel to direct our campaign. She has assisted organizations in raising close to $50 million. We accelerated our campaign's quiet phase due to our annual dinner in November so that our major donor efforts and community efforts are overlapping at this point.

Sustaining the Effort

We have a large base of dedicated funders who have been supporting HAIF housing operations for many years. There is an operational set-aside accounted for in the $16 million budget should any shortfall arise. However, we are being careful in our capital campaign to protect funding for operations and have no reason to expect difficulty in this area.

Features of the brand new HAIF Residence will be

- Located in a healthy neighborhood, modeling the attributes of a healthy family
- Emotional security, where safety and security will have a positive impact
- More common areas for classes and activities for both children and adults
- Separate rooms twice the size of those in the old facility
- Bathrooms for each family, enhancing privacy and quality of life
- Storage adequate for program needs
- Responsibly "green" design, thinking globally and acting locally

HAIF will strive for LEED Silver certification for building "green," assuming that it is appropriate financially given the priorities of the new facility.

We understand that a new building alone does not ensure the same or improved services for those who are being served. Ongoing operational expenses must be accounted for. With the physical benefits of our new location, we also have the opportunity to enhance and upgrade our staff and the transitional services we offer our families.

We thank you for your consideration, and look forward to your feedback.

Enclosures:
Case for Support
Board List and Comments
Campaign Steering Committee
501(c)(3) Letter
Financial Report
Annual Report
Statement of Nondiscrimination
Drawings of New Building
FAQs
DVD

Note to book readers: Project Cornerstone Asset Measurements are the 40 Developmental Assets people.

Sample Long Grant Proposal

A Proposal to the

We Want to Fund You Foundation

From The Best Aquarium in the Whole Wide World

In Support of General Operations

June 2010

The Best Aquarium in the Whole Wide World is deeply grateful to the We Want to Fund You Foundation for its generous past support. We turn to the foundation to respectfully request a renewed $15,000 grant to fund our most critical need, general operating support. Your gift will help to support the work of our entire organization, from mentoring programs for at-risk teens, to teacher professional development in public schools, to significant conservation initiatives in the Elephant Butte Reservoir region, to quality animal care and husbandry. The Best Aquarium in the Whole Wide World is committed to inspiring people to make a difference in their local and global communities through dynamic exhibits, education programs, and more.

Our multiple departments—Animal Health to Accounting, Education to Environmental Services, Facilities to Fishes—work in concert to bring the wonders of the aquatic world to the diverse audiences we serve. Guests view delicate leafy sea dragons and learn about the threats facing their wild counterparts in our Oceans Gallery; away from the aquarium, our Conservation Department works with international organizations, the Environmental Defense Fund, and Friends of the Earth to protect threatened habitats and to restrict destructive fishing practices. Our Guest Arrangement Department trains teens from economically depressed neighborhoods to teach our visitors about animal behaviors, whereas our Education Department connects teachers, students, families, and community organizations to enriching educational experiences. Behind the scenes, we share important health information about the animals we care for with members of the North American Breeding Cooperative and other zoos and aquariums. In our Regional Waters Gallery, visitors come face to face with invasive species that are threatening our lakes; our Elephant Butte Reservoir Region initiative brings these important issues into our communities through its comprehensive public relations campaign and partnerships. The oceans gave birth to our animals, but generous philanthropy sustains their life. Contributions in support of general operations from foundations such as the We Want to Fund You Foundation are integral to our success. We hope that you will collaborate with us in this important work.

The Best Aquarium in the Whole Wide World: History and Mission

As the largest indoor aquarium in the country, The Best Aquarium in the Whole Wide World celebrated its one-hundredth anniversary in 2000. Recognized nationally and internationally as a leading institution in its field, the aquarium features one of the largest, most varied collections of live aquatic animals in the world, comprising more than 35,500 fish, reptiles, amphibians, marine mammals, and birds and representing 2,000 species. Approximately 90 habitats offer windows into aquatic diversity of the continents of North and South America, Asia, Africa, and Australia and the regions of New Mexico and Central America. Each habitat, educational program, and publication underscores the fragile interdependencies of life and the local and global ramifications of every change in nature or action by human beings. "We are one planet, one people, and we care for multiple plant and animal species."

When we first opened in 1900, The Best Aquarium in the Whole Wide World was the only inland aquarium in the country that exhibited both freshwater and saltwater fishes.

In the spring of 1983, The Best Aquarium in the Whole Wide World unveiled its marine mammal pavilion and marine coastal area. This magnificent marine mammal pavilion was built on an additional 2.1 acres of landfill, and it doubled the size of the original aquarium building. Whales, dolphins, sea otters, and sea lions thrive in this dramatic recreation of a cold water coastal environment along with two bald eagles and four Coopers hawks.

To ensure our continued role as The Best Aquarium, our leadership developed a long-range Facility Strategic Plan in 1980 as an addendum to our overall mission-based Strategic Plan. The Facility Master Plan addresses exhibition, education, restoration, technology, and infrastructure needs, striving to ensure optimal care for the aquarium's animals while enhancing guest accessibility. The first entirely new exhibit created through this Master Plan was *Seasons of the River*. Opened to the public in June 2000, *Seasons of the River* enables guests to experience a year in the life of a South American ecosystem, learning through an interactive process how animals and human beings who live along the floodplain respond to the spectacular annual flood cycle. Last year, through a minicampaign, The Best Aquarium launched a new permanent exhibit that features the manatees of Florida.

Our mission statement reads: "At The Best Aquarium in the Whole Wide World, animals connect you to all living things and ecosystems so that you will become an advocate for our shared environment." Support from the We Want to Fund You Foundation will help to enable The Best Aquarium in the Whole Wide World to continue to foster this ethic by building audiences' appreciation, learning, and understanding of the aquatic systems in their own neighborhoods and beyond, strengthening science literacy, and encouraging direct conservation action.

Programs and Activities

Education

The Education Department offers programs that are designed to meet the needs of a wide-ranging audience, including Southwestern Public School teachers, children, teens, adults, and people in underserved communities. By working with advisory groups, including a Teachers' Council, and engaging in a project that asks indigenous community peoples to define their needs, The Best Aquarium in the Whole Wide World effectively connects its diverse audiences to the living world, inspiring them to make a difference.

Community Programs: Community programs seek to serve multigenerational audiences that are isolated from the aquarium and its resources because of various access barriers, which could include economics, language, culture, or previous museum experiences.

Following a community engagement strategy that focuses The Best Aquarium in the Whole Wide World's educational resources on targeted communities in the Southwest, the aquarium established community and neighborhood initiatives. These initiatives provide aquatic science programming, field trip experiences, and local stewardship opportunities to partner organizations serving southwestern residents. In 2006, we launched a neighborhood initiative in Paulson Park. Paulson Park is a community with more than 80 languages spoken in its households and represents a microcosm of the cultural and economic diversity of the Southwest. The neighborhood initiatives provide relevant, inspiring aquatic science and conservation information that the community then uses as it becomes a steward of its own local aquatic environments. The Best Aquarium in the Whole Wide World's Native Peoples Program, funded by a large federal agency, is an unprecedented community engagement plan involving native peoples and uses committees such as our Teachers' Council as models for involving other diverse groups of community stakeholders. These groups all work together in the development and implementation of The Best Aquarium in the Whole Wide World's education programs. The implications of this project are expected to be far-reaching as The Best Aquarium in the Whole Wide World commits to engaging its audience directly in the development of programming.

Mentor programs increase awareness of aquatic studies and careers among high school students in the Southwest. The Best Aquarium in the Whole Wide World's concept of mentoring is not a time-limited engagement with students; rather, the mentor programs staff strives to build lasting relationships with students. Mentor programs emphasize personal growth as well as professional exploration. As a bridge to this program, The Best Aquarium in the Whole Wide World staff developed a water experience on the Elephant Butte Reservoir Region to help students and their families, who come from a variety of cultural and socioeconomic backgrounds, to feel more comfortable in a sleep-away science experience away from home. As they participate in these programs, teens test ideas, ask questions, and seek advice from full-time aquarium employees who provide guidance and individual attention.

Student Programs: Student programs at The Best Aquarium in the Whole Wide World are committed to helping students and teachers build critical science skills and develop the confidence necessary to become competent practitioners of science as well as teachers of science through teacher professional development and inquiry-based student programs, all of which take place at school or at the aquarium.

Alongside student programs, the aquarium has a professional development program for teachers that is comprehensive and supports K through fifth grade teachers in their science teaching at 35 schools located in the Southwest. Its mission is to help participating teachers become more comfortable teaching hands-on, inquiry-based aquatic science through awareness, skills development, and use of best practices in science education. This program carries out its mission by giving teachers guidance tools that help them to increase their own science content knowledge, become more comfortable teaching science through inquiry methods, and use Aquarium resources in their teaching. These tools include orientations to the program, one-on-one mentoring, reduced-cost classes at the Aquarium, resource kits, and a curriculum that is tied to southwestern school standards and science sequence. Through these means, The Best Aquarium in the Whole Wide World endeavors to improve science teaching in public schools throughout the region.

Additionally, on-site classroom and laboratory programs engage teachers and students beyond the basic field trip experience. Programs complement the Aquarium visit and connect to southwestern states' learning standards. Each on-site class takes place in the Aquarium's Center for Education and builds on students' previous knowledge of aquatic science through "hands-on, minds-on" activities using unique Aquarium resources and environments that schools generally cannot provide. Last year, 94 percent of teachers surveyed rated their students' experience in a class or lab as excellent.

The Best Aquarium in the Whole Wide World's off-site programs are more popular than ever. Large group programs serve 75 to 250 students in grades K through 2, or 3 through 5 (up to 750 students per school) and are designed for students to become active participants in science inquiry and develop familiarity with the discipline. The Best Aquarium in the Whole Wide World's large group programs are presented at schools and use techniques that directly involve all audience members, enabling them to practice science, math, and even social science skills during the program. After each program, students receive free family passes to continue exploring at The Best Aquarium in the Whole Wide World. Along with the passes, the aquarium provides a guided tour through the Elephant Butte Reservoir Region for families. This is provided through a generous endowment from the We Want to Help Foundation.

Conservation

Elephant Butte Reservoir Region Initiative: Five years ago, The Best Aquarium in the Whole Wide World embarked on a groundbreaking Elephant Butte Reservoir Region awareness campaign that has reached thousands through a public relations campaign and through field restoration work that engages teens, young adults at risk, and senior citizens. The Aquarium is a logical rallying point for information on the Elephant Butte Reservoir Region, especially with its built-in audience that includes approximately 3 million adults and children annually and its close ties to the southwestern public school system. Given its prominence on the shores of the Elephant Butte Reservoir Region, it was logical that The Best Aquarium in the Whole Wide World assume this new leadership role.

The Elephant Butte Reservoir Region program also influences education programming as The Best Aquarium in the Whole Wide World's professional development workshops give teachers in the region the skills and resources they need to engage their students in inquiry-based science in the classroom through an Elephant Butte Reservoir Region–focused lesson that culminates in a two-mile hike and area cleanup. In the coming years, as we emerge as a center for the Elephant Butte Reservoir Region, The Best Aquarium in the Whole Wide World plans to invest further in its commitment through the development of a major permanent exhibit and complementary programming. Funding has already been secured through the Department of Justice and the Forest Service.

The Sustainable Lake Trout Program: A hallmark of The Best Aquarium in the Whole Wide World's conservation programming, this program is based on the belief that the power to restore healthy hatcheries lies with fishermen, consumers, and seafood purveyors. According to the journal *Science*, there may be a complete loss of lake trout in the Southwest by the year 2048 if we do not reverse our consumption trends immediately. The United Nations Food and Agriculture Organization estimates that more than 25 percent of the haul of fisheries worldwide is accidental wildlife caught in nets. While fish farming can be done responsibly, many aquaculture practices damage habitats or use ecologically destructive methods. Responding to this problem, The Best Aquarium in the Whole Wide World's Sustainable Lake Trout Program encourages consumers by providing accurate information on the numbers of fish spawned each season and providing them with recipes for alternative sources of protein.

Marine Mammal Breeding: Our expertise in aquatic science and animal health has led us to take a prominent role in actively breeding manatees—a first for any aquarium in the United States. In March 2008, Liza, one of the Aquarium's favorite manatees, gave birth to a healthy male calf. As part of the North American Breeding Cooperative, The Best Aquarium in the Whole Wide World shares the important information and knowledge gained from this rare birth with the marine mammal community. The Best Aquarium in the Whole Wide World's manatee breeding program and other public display programs have contributed to the recovery of these animals in the warm coastal regions of Florida and Bermuda. This population once exceeded 25,000 animals and then decreased to less than 5,000. Through intervention and new regulations based on research, our aquarium has contributed to the population's growth through research and artificial insemination techniques.

Collection and Exhibits Highlights

By connecting visitors to its collection, and thus to the living world, The Best Aquarium in the Whole Wide World demonstrates its commitment to both conservation and education. To best realize its mission, The Best Aquarium in the Whole Wide World puts its animals first, providing carefully designed habitats and exemplary animal care. From the halls of our animal hospital to the depths of our dolphin pool, our facilities are focused on the health and enrichment of our animals. This dedication to animal health makes the Aquarium a leader in its field, with accomplishments in microbiology, veterinary science, and animal husbandry. In addition, we continually evaluate the visitor experience and refine our exhibits to ensure that visitors have an engaging and inspirational experience with the wonders of the aquatic world.

Guest Arrangement has changed its focus from concentrating on animal encounters for guests to building a vibrant corps of more than 1,000 interpretive and guest service volunteers and teen field restorers who will contribute to an impactful guest experience. This department also will pilot a new workforce development teen program supporting the Guest Services Division. Guest Arrangement will implement a comprehensive brand-focused training program that will raise the bar of guest satisfaction for the Aquarium's public face.

Supporting The Best Aquarium in the Whole Wide World's mission to connect guests to all living things and ecosystems so that they will become an advocate for the environment, Guest Services and Animal Welfare now offer a variety of engaging educational experiences that take place in the aquarium's galleries and special exhibits. Floor programming, where guests have an opportunity to "meet the collection," reaches the majority of the more than 3 million annual visitors, directly connecting all—from families to adults to school groups—to the Aquarium's diverse collection representing life in the earth's oceans, lakes, rivers, and streams. From the Marine Mammal Pavilion and Marine Coastal Area's presentation that reaches thousands of guests at a time to an intimate animal encounter that encourages a special one-on-one interaction to our Seniors on Saturday program serving our oldest (but highly visible) visitors, educational experiences in the Aquarium galleries and exhibits are often the highlight of our guests' visit, inspiring a connection to nature and promoting conservation awareness and encouraging conservation actions.

Two of our most popular Guest Arrangement offerings include the Polynesian Reef Dive Presentation, which affords visitors the unique opportunity to converse with a diver while he or she is feeding animals within the Polynesian Reef Exhibit, and the Marine Mammal Presentation, which features our manatees, whales, and dolphins. Combining natural history, conservation, and animal care information, The Best Aquarium in the Whole Wide World's presentation programs provide meaningful insight into the animals and exhibits at the Aquarium. We also offer a private animal encounter program that connects guests to animals in the Aquarium's collection through beyond-the-exhibit interactions.

Moving away from traditional gallery interpretation, The Best Aquarium in the Whole Wide World recognizes that experience is the most effective delivery mechanism for information. For example, an innovative program that combines play experiences, gallery interpretation, and animal encounters in an exhibit-quality setting is called the *Campground Playscape at the Elephant Butte Reservoir*, located in one gallery. Developed in 2007, the *Playscape* debuted in September 2008 and offers a complete interpretive experience for guests: a canoe interactive simulating a leisurely ride down the Elephant Butte Reservoir River, a campsite outfitted with a tent and camping equipment, recorded songs of native peoples to which children can sing along, and an animal encounter stage with waterfall, turtle pond, simulated rockwork and bird of prey roost.

Evaluation: Part of The Best Aquarium in the Whole Wide World's success lies in its stringent commitment to ensuring that every exhibit, conservation initiative, and education program fits into our overall mission. The Best Aquarium in the Whole Wide World has on staff five full-time evaluators in its Audience Research area who conduct yearly studies concerning member services, design, exhibits, education programming, and guest services. As with past programs, focus groups, attendance figures, surveys, and requests from current program users will give us front-end evaluation, while staff, outside evaluation consultants, and advisory groups will assist in the evaluation of program structure and content.

A gift of $15,000 from the We Want to Fund You Foundation will help The Best Aquarium in the Whole Wide World to continue sharing the wonders of the aquatic world with communities throughout the Southwest and the Elephant Butte Reservoir Region, inspiring millions each year to connect with the environment and keep it protected. Thank you for your consideration of this request.

17

Program and Project Budgets

The following examples show different ways of presenting a budget with your grant proposals. Three different types of projects are presented in three different formats to demonstrate the flexibility you have to exhibit your budget. Choose any format that clearly shows how you will use the funds. Add any other sources of funds for the project, and if necessary, include notes that clarify budget elements. Grantors know when a budget is overinflated, so give as accurate an estimate as possible. Be careful not to underestimate either. Your organization will be on the hook to complete the project with the amount of funds you requested. Sometimes a grant application is restrictive as to word counts and doesn't allow room for budget notes. Slip those words onto the budget page. (Relating them to the budget gives you an extra 50 or so words).

Don't forget to add your organization's name to the budget page (and all your attachments). Since you went through the effort of selecting an awesome project title, don't forget to include it. It reminds the readers of the project they enjoyed reading about.

The ABC Aquarium sample shows both the education program total budget and the proposed project budget on one page—note the "budget notes" that remind the reviewer of the project—while the Dinosaur Dreams project budget includes remarks on how the budget figures were arrived at. Depending on the complexity of your project's budget, you may need notes.

Be clear, be clear, be clear! Many a grant proposal has been denied because of an ambiguous budget. And finally, make certain that the figures add up correctly!

Sample Budget

ABC Aquarium

Education Program Budget 2011–2012

Volunteer Program	$3,500
Teen Mentoring Program	$1,000
Supplies and Materials	$10,100
Education Program Staff	$96,500
Training and Development	$2,000
Travel	$1,500
Total	$114,600

ABC Aquarium

Education Program Enhancement Project Budget

Education specialist certification	$1,000
Training for education program enhancement	$1,500
Travel to curriculum development training	$650
Total grant request	$3,150

Budget Notes: XYZ Council's support will help to fund the cost of providing unique ocean science educational classes, including curriculum development, implementation, and evaluation.

The project includes fees for the education specialist to attend a regional conference and obtain an Aquarium science teaching certification. ABC Aquarium will cover the food and lodging expenses, estimated at $400.

Sample Budget

YYZ High School

Football Field and Track Project Budget

	Request to ABC Foundation	Confirmed Support from EFG Corporation	Confirmed Support from XYZ High School Parents Foundation	Total
Bleacher refurbish	$1,000		$5,000	$ 6,000
Track surface		$40,000		$40,000
Field lighting	$5,000			$ 5,000
Scoreboard	$5,000			$ 5,000
Total	**$11,000**	$40,000	$5,000	$56,000

Grant request to ABC Foundation	$11,000
Other sources of funding	$45,000
Total project cost	$56,000

Sample Budget

Natural History Museum

Dinosaur Dreams Project Budget

Projected Revenues:	
ABC Foundation	$250,000
XYZ Foundation	$200,000
XYZ Council	$100,000
City of Seattle	$ 50,000
Other sources (individuals)	$ 50,000
Total project revenues	**$650,000**
Projected Expenses:	
Architectural services (10% of construction cost)	$ 31,500
Construction (exhibit hall renovation @ $90/sq ft)	$315,000*
Cast concrete dinosaur skeletons and installation	$180,000*
Project and construction management	$ 32,500
Interpretive signage	$ 9,025
Interactive features	$ 30,000*
Contingency (15% of architectural and construction cost)	$ 51,975
Total project expenses	**$650,000**

*Detailed budgets of these line items are available on request.

Appendix A

List of Items Available on Request

Sample List of Items Available

The Items Available Upon Request document is a sample of the types of items you may want to make available to potential grantors upon the submission of a grant proposal. You might use this list in cases where: the funder does not allow attachments, you're submitting a short proposal or a query letter, the funder has an extensive application but it does not ask about some elements about which you want them to know. For example, your organization may have an awesome annual report with photos of your key audience, great descriptions of your many programs, and last year's record-breaking statistics. Or, if you think it strengthens your proposal for the grantor to see the biographies or curriculum vitae of key staff—but there wasn't room in the proposal—offer to send the items by including them on your list.

This list is optional. Many proposals will not need one. Also, not all of these items may suit every proposal, so you can adapt it as necessary. Possibly you do not have all of your documents ready to submit. In that case, do not include them on the list. A grantor could call you and ask you to e-mail an item, so only list those that exist and are ready to share.

There are two ways to offer this list. First, as an attachment. Identify your attachment(s) below your signature, after the closing paragraph. Second, if attachments are not allowed, add the list in the body of the proposal, just before the last paragraph, for example, "Items available upon request include: annual report(s), photos, newsletter, and naming opportunities." Then go on to your closing paragraph.

You could also identify any one item in the body of your proposal as the subject arises. For example, in the Program Background section, you may want to end the discussion with a statement, in parentheses: "(Key staff biographies are available upon request)." Or for a Community Collaborations question in your proposal, you may want to add this statement to your response: "(Letters of Recommendation are available upon request)." Then in these cases, you wouldn't need the list.

Even if a funder allows attachments, a list like this saves postage and printing costs by not sending them things they have no interest in seeing. The comprehensive list below is a sample of the many items organizations may have. But not all of them are required.

Also, if a funder has never heard of your organization before, it may reassure them to see that your organization has some of these items, for instance a strategic plan, even if they don't request to see it. Including mention of important pieces, whether in the body of the proposal or as an attachment, reinforces your grant proposal. It implies a sense of stability and professionalism.

Your Organization Name

Items Available on Request

501(c)(3) Letter
Annual Report(s)
Awards and Recognition
Community Partnerships
Key Staff Biographies
Key Staff Curriculum Vitae
Letters of Recommendation
List of Board of Directors
Naming Opportunities
Newsletter(s)
Operating Budget
Organization History
Photos
Program Brochures
Schematic Drawings
Strategic Plan

Please call or e-mail [Your Name] to request any of these items:
(xxx) xxx-xxxx or xxxxxx@xxx.com

Appendix B

List of Ways to Earn Funds

As mentioned in Chapter 7 and Chapter 10, your organization ultimately will want to consider other ways to earn funds. By not relying on just one form of fundraising, that is, grants from corporations, your organization will be better able to handle the blow if your corporate funders change their grant programs to alter or end your ability to request support from them. Combining corporate grant writing with foundation grant writing helps, but it's healthier to find other methods entirely to raise funds, raise awareness, and raise community support for your organization. All three of these are indispensable for the endurance of your organization.

Ask others in your organization to join in the brainstorming to discover unique, affordable, and doable ways for your organization to earn funds. Once you have your own list of potential ways to earn funds, pick the one you'll explore further. Does it reinforce your mission? Are there other successful models of this method (or similar to it) to look at? Does it create too much work for staff? Is it costly to implement? Is it a one-time event, or does it have long-term potential? Basically ask, "Is this a good idea for us?"

In the following sample, remarks at the bottom of the page remind everyone where their income currently comes from and what they've tried in the past.

Remember that the purpose of fundraising is to raise funds, raise awareness of your organization, and raise community support. If the new idea also can reinvigorate staff and board, generate media buzz, or create new partnerships, all the better!

Sample List

Ways to Earn Funds for XYZ Senior Center and Park

Grant applications to foundations

Grant applications to corporations

Corporate sponsorships

Naming opportunities (buildings, rooms, benches, fountains)

Donations from individuals (major gifts)

Employee corporate matching funds

Workplace giving annual campaigns (such as United Way)

In-kind goods or services

Will bequests

Annual fund drive

Visa/Mastercard royalty

Online donations through Web site

Social media fundraising

Auction (online or on-site)

Donor tiles or bricks

House party

Established revenue sources include membership dues, admission fees, exhibit donation boxes, and interest on funds. Past fundraising included raffle drawings and an annual gala dinner. We also received a three-year arts grant from the state arts commission.

Appendix C

The Best Online Grant Writing Resources

GrantWhisperer.com

This is my Web site. It offers more grant writing tips, inspiration, and grant writing resources. Visitors also have the opportunity to ask me questions and join in the discussion.

Foundationcenter.org

This site offers a wealth of grant writing information that includes free podcasts of funder's presentations and a Foundation Finder search tool that can help you research potential foundation funders. The 990 Finder is another valuable free tool to research past grants awarded by a specific foundation. Under the "Get Started" tab, click and visit the FAQs page for answers to many proposal-writing questions. The Foundation Center has five locations, so click on the city nearest you for local offerings (some workshops are free to you). For a monthly fee, you can browse through the online foundation directory and view any foundation's information page or the corporate giving database and conduct an online search for corporate donors.

Foundationcenter.org/getstarted/learnabout/proposalwriting.html

This page lists the free online courses in proposal writing (in several languages) and a free proposal writing Webinar.

Tgci.com

The Grantsmanship Center has much to offer. For example, under the "Resources" tab you'll find a free e-newsletter, *Centered*, with short fundraising articles; select "Magazine Archives," and you'll find an abundance of proposal-writing articles. Under the "Funding Sources" tab, click on your state, and then choose between foundation searches or corporate giving for that state. All the foundation pages have active links with valuable information and links to the foundation Web sites.

Guidestar.org

Guidestar has a free search function if you know at least part of the foundation's name. On the first page, under "Search Guidestar," there is a field that says "Nonprofit search." Type in the foundation's name, and click "Start search." Here's the thing: You also can type in "Foundation" plus your state and see what pops up. For example, "Foundation, Ca" yields over 35,000 names of foundations. (Not too helpful because the bulk of those are entities like you, seeking funds.) However, if I type, "Foundation, Ca" and my organization's zip code, only 36 names pop up. This is a much more manageable number. Plus, each listing has a one-line description so that you can tell immediately whether its purpose is to seek funds or give funds. (Also check the zip codes surrounding your organization.) You'll need to register (free) to search your selected foundation's Form 990(s). As you know, with the Form 990, you can find valuable past funding information, address and telephone number, and the foundation's contact person and its board members.

Foundations.org

This site is provided by the Northern California Community Foundation, Inc. However it has two free national databases. From the home page, click on "Directories," and then click on "Research Corporate and Private Foundations" and "Grantmakers Directory" to get to a directory with live links to corporate grant makers. Or you can bypass the home page and use the direct URL: www.foundations.org/grantmakers.html.

From there, you can click on "Community Foundations" or, from the home page, click on "Research Community Foundations Directory." This is a slim listing, but you may find a useful match. To bypass the home page, use the direct URL: www.foundations.org/communityfoundations.html.

While the home page may feature a foundation or other grant opportunities, be aware of the fundraising ads with enticing titles.

Grants.gov

I briefly discussed this site in Chapter 6, but there is more to this site to point out. From the home page, click on "Find Grant Opportunities," and then click on "Advanced Search." The Advanced Search page allows you to narrow your search by several factors, such as eligibility, agency, and funding category. This page also has a link to a PDF "Guide to Searching Grant Opportunities," which will help you to navigate through using the site's search options. At this time, the direct URL for the advanced search page is www07.grants.gov/search/advanced.do.

I'd also recommend going to the agency's Web site directly. For example, the U.S. Department of Education home page has a "Grant Opportunities" tab, the National Science Foundation home page has a "Funding Opportunities" tab, and the U.S. Department of Health and Human Services home page has a "Grants/Funding" tab that has a link to "Tips for Preparing Grant Proposals." Some of these agency sites give grant samples and guidelines for their grant opportunities.

www.Imls.Gov/Applicants/Sample.Shtm

The Institute of Museums and Library Services Web page for "Grant Applicants" deserves special attention. The page provides several grant proposal samples of various funding categories and award amounts. This is a very user-friendly site that I recommend to anyone who wishes to apply for a federal grant regardless of the agency you plan to target. The tools and materials will help to demystify federal grant applications.

Nea.gov

The National Endowment for the Arts is another federal agency with a great Web site for grant seekers. From the home page, click on "Grants" to be taken to a simplified site that will lead you to forget it's a government agency. It is easy to find recent grant awards with nice descriptions that may help you to describe your proposal project. There are good tips and a nice artistic drawing of the life cycle of an application.

Researchguides.library.wisc.edu/content.php?pid=16143&sid=108666

The University of Wisconsin–Madison "Libraries" page has numerous useful online resources for grant writers. Three focus areas include links for federal funding, nongovernmental funding proposals, and research funding proposals. Many of the links provide grant proposal samples.

www.search-institute.org/grant-writing-resources

The Search Institute offers a page with grant writing resources. (You may remember that this organization is the 40 Developmental Assets group.) The institute focuses on those serving youth as well as educators; however, anyone might find the sources useful.

Npguides.org

The Nonprofit Guides Web site provides grant writing tips and a sample letter of inquiry and includes four full sample grant proposals.

Grantproposal.com

This Web site offers sound advice, tips, and samples for grant writers. It even provides a list of active verbs to use in your grant proposals.

www.Purdue.Edu/Envirosoft/Grants/Src/Msieopen.Htm

For something a little bit different, visit this site created by Purdue University for the Environmental Protection Agency (EPA). While the focus is on environmental grants, there's so much good information, it's worth a look. Click on the items on the left border to go to that page. I particularly like the "Enhancing a Proposal" tab. Each component of the sample is described briefly. The "Examples" tab will show you full proposals such as Butterfly Gardens in Schools and Birds Without Borders. There's also a glossary of environmental-related terms.

Schoolgrants.org

The page at this site that will interest grant writers is www.k12grants.org/samples/. This page offers an index of grant samples aimed at educators. These samples are available free of charge and many have resulted in hundreds of thousands of dollars in funding. You'll also find useful grant writing tips.

Afpnet.org

The Web site of the Association of Fundraising Professionals (AFP) includes fundraising news and reports that are available to anyone. Some features are available only to members. If you are writing grants regularly or taking on the fund development role at your organization, seriously consider joining AFP, which includes membership in your local chapter. Local networking and learning may prove invaluable to your organization.

Other grant writing organizations include the American Association of Grant Professionals. Their Web site is grantprofessionals.org, and the Northwest Development Officers Association is at ndoa.org.

Index

About the Author

Victoria M. Johnson, CFRE, has been a successful grant writer for over 10 years at a northern California zoo. Her professional and volunteer work led her to write this book to help all those friends who ask her about grant writing. She blogs at grantwhisperer.com to further help those in need of grant writing tips, inspiration, and an occasional kick in the pants. She's an energetic speaker at workshops and conferences, and she interviews fundraising experts on her "blogtalk" radio show at blogtalkradio.com/grantwhisperer. She truly believes that we can change the world, one grant at a time.

CPSIA information can be obtained
at www.ICGtesting.com
Printed in the USA
BVOW11s1344090717
488862BV00010B/54/P

9 780071 750189